Merry Christmas
to Connie
Love Patty 1980

Le Barbier lainé Inv. *N. le Mire Sculp. 1783.*

'Rousseau the Herbalist', an engraving by N. Lemire after a painting by Le Barbier. (*Bibliothèque Nationale*)

On the previous page: A silhouette of Rousseau (*Bibliothèque Nationale*)

PURE CURIOSITY

Botanical Letters
and
Notes towards a Dictionary of Botanical Terms

by
JEAN-JACQUES ROUSSEAU
Illustrated by P.J. REDOUTÉ

Translation of the Letters by Kate Ottevanger

PADDINGTON PRESS LTD
NEW YORK & LONDON

Library of Congress Cataloging in Publication Data

Rousseau, Jean-Jacques, 1712-1778.
Pure Curiosity

Translation of La botanique de J.-J. Rousseau.
Bibliography: p.
1. Botany – Pictorial works. I. Redouté, Pierre Joseph, 1759-1840. II. Title.
OK98.R613 580 79-13253

ISBN 0-448-22920-X

Filmset in England by Pierson LeVesley Limited
Printed in Switzerland by Hallwag AG Bern
Bound in Switzerland by Maurice Busenhart, Lausanne
Color separation by Hallwag AG Bern
Editor: Robin Wright
Art Editor: Shelagh McGee

In the United States
PADDINGTON PRESS
Distributed by
GROSSET & DUNLAP

In Canada
Distributed by
RANDOM HOUSE OF CANADA LTD.

List of Contents

Botanical Contents of the 'Elementary' Letters

The First Letter: *The true use of botany – importance of learning the structure of plants, not merely their names – component parts of a plant – explanation of reproduction/fructification – the parts of the flower: corolla, pistil and stamen as exemplified in the Lily – character of the Lily family – botany a study of observations and facts.*

The Second Letter: *Double flowers to be avoided in botanical examinations – analysis of the Wall-flower as an example of the crucifer – division into two orders, siliquose and siliculose – small flowers requiring examination with a magnifying glass – other botanical instruments.*

The Third Letter: *Botany to be studied from nature, not books – analysis of the Pea flower – regular and irregular flowers – protection of fructifying processes in leguminous or papilionaceous family.*

The Fourth Letter: *Reason why in crucifers two stamens are shorter than the other four – glands at the base of the filaments – mouth-shaped (labiate) flowers – analysis of White Dead-Nettle – personate flowers as exemplified in Snapdragon and Toadflax – comparison of labiate and personate flowers – true way of distinguishing them.*

The Fifth Letter: *In cruciform flowers glands are very small – botany not a science of words, but one which teaches the structure of plants – arrangement of the inflorescence in the umbelliferous family – division of flowers into inferior and superior – description of the flower and fruit of umbellifers – true character of the family – how to avoid mistakes in ascertaining this character, especially in Elder and Eryngium – great similarity of all umbellifers – difference between Hemlock, Chervil and Parsley.*

The Sixth Letter: *Umbellifers and other genera to be known by their general appearance and analysis of inflorescence – structure of composite flowers exemplified in the Daisy – different kinds of florets found in composite flowers – receptacle an important part of the composite flower, as seen in the Dandelion – the calyx – structure of the floret and the demi-floret – the use of tuft to seeds – change in the form of the calyx – flowers well adapted to examination.*

The Seventh Letter: *Botany, a study of pure curiosity – nature frequently disfigured by man – to appreciate nature look in forests, not kitchen gardens – fruit trees, though grafted, retain their botanical characteristics – different fruits are but varieties – Pear, Apple, Quince, Cherry, Plum, Apricot, Almond, Peach, Nectarine, their characteristics and that of the family to which they all belong.*

The Eighth Letter: *The dried plant collection – its use, to remind us of plants we know, not to give us knowledge of plants we have never seen before – how to form a collection of dried plants.*

INTRODUCTION

In the spring of 1878 John Ruskin, having temporarily recovered both his sanity and his childhood passion for plants after a winter of madness, sent off an imperative note to the bookseller F.S. Ellis:

> Please at once set your Paris agents to look out for all copies that come up, at any sale, of Rousseau's *Botanique* with coloured plates, 1805—and buy all they can get.

Although the agents presumably did their best, they did not find a single copy. And Ellis may have smiled at the optimism in Ruskin's order, for by 1878 the work in question had long since vanished from auction rooms into the cabinets of learned institutions and private collectors. In fact, it had been recognized on publication as an item for connoisseurs of flowers and books, so much so that a few weeks later in 1805 there had been a second edition and then in 1821, still with the original plates, the handsome (and now also rare) third edition largely reproduced in the present volume.

Period sensibility was one of the reasons. During the eighteenth century and the early nineteenth botany was fashionable: the inquiring layman who in our post-Einstein era is attracted by quasars and the big bang was in those post-Linnaeus decades interested in stamens and pistils. Indeed, 'interested' is scarcely the word, for there were moments when botany seemed about to become a new religion, marked by adoration of Flora, by ritual outings, by pilgrimages to Kew Gardens or the Paris Jardin des Plantes, and by the incantatory use of the Latin names for genera and species. Statesmen, duchesses, bluestockings, and thousands of ordinary citizens compiled herbaria. Fortunes went into exotic plants: Joséphine Bonaparte died with debts of more than two million francs, mostly because of her addiction to botany at Malmaison.

Recalling that scented, bygone mania can add to the enjoyment of this book. But a Zeitgeist cannot, of course, automatically turn out a wonderful flower album, and so one should also recall that here there was brilliant help from Jean-Jacques Rousseau, Pierre-Joseph Redouté, and a team of engravers and printers.

Rousseau remarks in his *Confessions* that he might have become a great botanist, 'for I know of no study in the world that suits my natural tastes better than the study of plants'. As early as 1738, when he was twenty-six and still living in the country near Chambéry, he was tempted by 'the contentment . . . in the eyes' of an acquaintance returning from a herb-gathering expedition. But then came the years in Paris, occupied by music, politics, and love affairs; by stormy relations with Voltaire, Diderot, and Grimm; by the writing of *La Nouvelle Héloïse*, *Le Contrat social*, and *Émile*; and eventually by a chronic fear of the policeman's rap on the door. The Jean-Jacques who at last took up botany was a hunted, middle-aged neurotic who wanted to escape from everything and everybody, himself included.

In 1762 his *Émile* was condemned by the Paris Parlement, and he was threatened with imprisonment. He fled to his native Switzerland and sought refuge at Yverdon, which was then under the control of Bern. Chased out by the Bernese authorities a month later, he moved north into the territory of Neuchâtel and installed himself at Môtiers, where he was soon botanizing regularly. In a letter of 1764 he asks for information about a recent book on plants, and adds a bit of humility:

> I should like very much to subscribe for this publication, since like a true dotard returning to childhood I am trying furiously to learn botany without the guidance of a single book. With the help of this study my solitary strolls, in a countryside as rich in plants as this one is, can be delicious.

Apparently he rapidly acquired the books he needed, and perhaps his humility was feigned, for at about this date he turned encyclopaedist and began to plan a *Dictionnaire des termes d'usage en botanique*.

Although the project was finally abandoned, the completed introduction affords a good view of a man of the Enlightenment immersed in methodology. He begins with a plea for pure science:

> The first misfortune of botany was to have been regarded at birth as a branch of medicine. The result was that people concentrated on finding, or imagining, the virtues of plants, and neglected knowledge of plants themselves.

At the close of the Middle Ages, we are told, this first misfortune was followed by another, that of bookishness: 'Plants were no longer studied on the ground, but in Pliny and Dioscorides.' In the end, however, adventurous spirits began 'to seek, to observe, to conjecture' and thus to find themselves up against the problem of nomenclature, which is what Rousseau wants to argue about. He reviews some of the systems still in use in the first half of the eighteenth century, and deplores their long-windedness and their bad effect on education:

> Nothing was more depressing and more ridiculous, when a woman or one of those men who resemble women asked you the name of a plant or a flower in a garden, than the obligation to answer by spitting out a string of Latin words that resembled a magic invocation. Such a pedantic apparatus was enough to make these shallow-minded persons recoil from a fascinating study.

But the remedy is at hand. The author has already expressed strong approval for the relatively simple binomial nomenclature developed by Linnaeus, whose monumental *Species Plantarum*, containing descriptions of some six thousand species of plants from many climates, had appeared in 1753:

> At last Monsieur Linnaeus, full of his sexual system and of the vast ideas it had suggested to him, formed the project of a general reorganization, which everyone felt was needed but which nobody dared to undertake. He did more than form the project; he executed it. . . . In doing so he had to create, so to speak, a new language for botany.

Such a language, Rousseau concludes, is as essential for the study of plants as algebra is for geometry.

A series of disasters interrupted his dictionary-making. He was savaged in an anonymous pamphlet (actually written by Voltaire); his *Lettres écrites de la montagne* produced a great outcry in Geneva. The Protestant ministers of Môtiers turned violently hostile. His house was vandalized during the summer of 1765. In September he retreated with his housekeeper and mistress, the ex-chambermaid Thérèse Levasseur, to the tiny island of Saint-Pierre, in the Lake of Biel; and there, as related in one of the most Romantic passages of the *Confessions*, he spent six happy weeks solacing himself with Linnaeus and the local flora:

> Taking the entire island as my botanical garden, whenever I needed to make or verify some observation I would run through the woods or across the meadows, my book under my arm; then I would lie prone on the ground next to the plant in question, in order to comfortably examine it on its site. This method was of great help to me in getting to know plants in their natural state, as they are before they have been cultivated and denatured by the hand of man. . . . Sometimes I would cry out with tears in my voice: O nature! O my mother! . . . With pomp, we brought in rabbits to populate [a neighbouring islet]: another fête for Jean-Jacques.

A contemporary portrait of Madame Delessert after a painting by M.H. de Rothschild, from the Bartholdi Collection. (*Librairie Armand Colin, Paris*)

Expelled from this Eden at the end of October, he moved on to Biel, to Strasbourg, to London, to Chiswick. Richard Davenport, a rich merchant with sympathy for eccentric genius, offered the use of Wootton Hall, a pleasantly rural seat on the Staffordshire-Derbyshire border near Dovedale; here the fugitive devoted the summer of 1766 and the following winter to writing the first part of the *Confessions*, to exhibiting increasingly clear symptoms of paranoia, and to gathering English wild plants—in particular the ferns and mosses of the neighbourhood. After quarrelling absurdly with David Hume he returned in panic to the Continent, and passed the next three years wandering through half a dozen French provincial towns. In the summer of 1770 he resumed his old trade as a copyist of musical scores and settled again in Paris, the police having mercifully hinted that they would ignore his presence.

He took a worker's apartment near Les Halles in the rue Plâtrière, at what is now 52 rue Jean-Jacques Rousseau. His mind was dark with the suspicion that he was the victim of a universal plot; and his days, as described by his friend Bernardin de Saint-Pierre, were often more botanical than musical:

> In the summertime he got up at five in the morning and copied music until seven-thirty. Then he had breakfast, during which he busied himself with arranging on paper the plants he had gathered the previous afternoon. After breakfast he went back to copying music. He dined at twelve-thirty. At one-thirty he often had a coffee at the Café des Champs-Elysées . . . Then he went out into the countryside to collect plants, his hat under his arm in the hot sun. . . . He maintained that the action of the sun's rays did him good.

In 1766 he had begun to discuss botany with a number of correspondents, including notably the Duchess of Portland. Now, in the rue Plâtrière between 1771 and 1773, he wrote the eight 'elementary' letters that appear in this book.

These were addressed to the young Mme Étienne Delessert (1747-1816), née Madeleine-Catherine Boy de La Tour and married to a prosperous banker at Lyons. Rousseau had long known her family in Switzerland; and at Môtiers in 1762, when she was fifteen, he had earned her gratitude by helping to rescue her from a projected and very unromantic marriage of convenience. He had acquired the habit of referring to her, fondly and familiarly, as Madelon or, although they were not related, as *cousine* (roughly with the connotations of the archaic English 'coz'), and had extended his affection to her sisters, Julie and Élisabeth, and to her widowed mother. Her intelligence and judgement, he felt, were 'far greater than her age and equal to the excellence of her heart'. She had responded by writing to him regularly during his years of wandering, by tolerating his uncertain moods, and by providing him with a small harpsichord;

and in 1771 she had perhaps pleased him most by asking for his help in the botanical education of her four-year-old daughter, Marguerite-Madeleine. (One can suppose that the child was very bright, or that Mme Delessert intended to carry out Rousseau's suggestions a little later on.)

In the first letter of the series the preceptor remarks slyly that he is afraid of playing the role of Molière's Monsieur Josse, who prescribes according to a vested interest. Then he moves into a practical application of the views expressed in the introduction to his abandoned dictionary. Marguerite-Madeleine is to begin by learning to 'really see what she is looking at'; like a good little disciple of Linnaeus, she must scrutinize the parts, structures, organization, and sexual characteristics of plants. Thus, almost insensibly, she will acquire the rudiments of a sound system of classification and the framework for binomial nomenclature. The educational method is not far from what is advocated in *Émile*, and there are other echoes of the main tenets of Rousseau-ism: in the seventh letter, for example, civilized man is scolded once again for having meddled with nature, and 'those double flowers admired in flower-beds' are dismissed as 'monsters deprived of the faculty of reproducing their kind'. (In his rejection of such 'meddling' Rousseau, faithful follower of Linnaeus and believing like him in the fixity of species, is touching on the growing debate on transformism ('the hypothesis that existing species are a product of the gradual transformation of other forms of living beings' OED). He was certainly aware of the contemporary discussions and experiments which were to prove evolution, but ultimately rejected the whole debate because of his own religious views.)

But the pedagogy is saturated with the tenderness that emerges in the allusions to *la petite* and in the *bonjour, chère cousine*; and statements of doctrine are accompanied by a lightness of touch, a sort of gentlemanly amateurism, that at times comes close, although this was certainly not the intention, to a discounting of the value of science itself:

> You must not, dear friend, give botany an importance it does not have; it is a study of pure curiosity, one that has no real utility except what a thinking, sensitive human being can draw from observing nature and the marvels of the universe.

An unalerted reader of these charming lectures would be unlikely to suspect that their author was a sufferer, with delusions of persecution (which were not, of course, entirely delusions), who desperately needed the consolation of a discipline that could make its devotees forget the drab utility and the wickedness of modern civilization. Clearly, botany was not always just 'a study of pure curiosity'. On occasions it became a quest for knowledge that engaged all of his considerable intellectual energy.

There were times when the quest exasperated him. In July 1776, in a letter to the Duchess of Portland, he says that he has rid himself of all his books on botany, for 'the agreeable amusement' has become 'too tiring'. But by the seventh 'promenade' of his *Rêveries du promeneur solitaire*, written in 1777 less than a year before his death, he has changed his mind completely:

> Suddenly, past the age of sixty-five, deprived of the small memory I had and of the remaining strength I had for running across the countryside, without guide, without books, without garden, without herbarium, here I am caught again in that mania, with even more ardour than when I succumbed to it the first time . . .

He forms 'the sage project' of learning all of Linnaeus by heart; he plans herbaria that will contain 'all the plants of the sea and of the Alps and all the trees of the Indies'. His botanizing seems to him justified by a comparison of the study of plants with other sciences. 'The mineral kingdom', he decides, 'has nothing that in itself is pleasant . . .' As for the study of animal life, it has the defect of leading to the dissection room:

> blood, disgusting intestines, frightful skeletons, and pestilential vapours! It is not there, on my word, that Jean-Jacques will go looking for amusement. Come bright flowers, variegated hues of the meadows, cool shade, small streams, groves, greenery, come and purify my imagination, soiled by all these hideous objects.

He continued, on through his last weeks of strolling in the 'philosophical' park of Ermenonville, forty kilometres north of Paris, to be consoled by botany.

Although his letters on the subject seem to have circulated a good deal in intellectual salons, they remained hidden from the general public until their appearance at Geneva in 1781 in a purportedly complete collection of his work. Here they caught the eye of Thomas Martyn, professor of botany at the University of Cambridge, who in 1785 brought out an English translation of the eight letters to Mme Delessert and the dictionary introduction, accompanied by twenty-four additional 'letters' written by himself somewhat in the manner of Rousseau. French versions came out in Paris in 1800 and 1802. And then Redouté altered the situation in a spectacular way by participating in the preparation of the 1805 edition that Ruskin would try to buy.

Over the last two hundred years there have obviously been advances in botany and changes in terminology but as far as the letters are concerned these are remarkably few.

An engraving by C.S. Pradier after a drawing by Gerard.
(*Bibliothèque Nationale*)

The Raphael of flowers

The life of Redouté is proof that plants, skilfully and lovingly depicted, can bring a man safely to port through almost any storm history can blow up. He was born in 1759 at Saint-Hubert, a village in a part of the Ardennes that then belonged to the Duchy of Luxembourg and now belongs to Belgium. His father painted, and so eventually did two brothers. At the age of thirteen, having learned what he could at home, he set out with his paintbox to explore Flanders and the Low Countries. During the next decade he earned an uncertain living as an itinerant artist (in the same tradition as that of the 'limners' who were roaming the American colonies), studied the available old masters, managed to acquire a year of further training in an atelier at Liège, and underwent something like a conversion, in Amsterdam, by discovering the work of the finest of the Dutch eighteenth-century flower painters, Jan van Huysum. In the meantime the eldest of the Redouté brothers, Antoine-Ferdinand, had become a stage-set designer in Paris, and in 1782 Pierre-Joseph followed suit. By now, however, his first interest was botanical illustration. He began to frequent the Jardin des Plantes (then still the Jardin du Roi but already, under the energetic Buffon's direction, a splendid scientific enterprise), where he met a wealthy Linnaeus disciple, Charles Louis L'Héritier de Brutelle, and profited from the encouragement of Gerard van Spaëndonck, a Dutch water-colourist who had become the institution's professor of flower painting. L'Héritier took the obviously gifted young provincial under his wing, gave him access to a large botanical library, and involved him in a trip to England to study the exotic plants at Kew. By 1788 Redouté was the illustrator of two of his patron's books, *Stirpes Novae* and *Sertum Angicum*; a year later he was named, probably at L'Héritier's suggestion, draughtsman to the cabinet of Marie-Antoinette. He was on his way.

People, plant-lovers included, closely linked to the *ancien régime* had little reason to expect favours during the Revolution; Malesherbes, for example, was guillotined, and L'Héritier was imprisoned. Moreover, in 1792 Redouté worsened his case by yielding to a request from the bored Marie-Antoinette, then confined to the donjon of the Temple, to come and paint a cactus of which she was fond. Yet he survived in excellent shape. During the Terror he was appointed to the staff of the former royal botanical garden, which had become the Jardin des Plantes and the Muséum National d'Histoire Naturelle, with the Rousseauist Bernardin de Saint-Pierre as director. By now he was showing his pictures of flowers, fruit, and mushrooms in the official Salon and serving on the powerful jury of well-known painters (including David, Vien, Gérard, Fragonard, and Carl Vernet) that dispensed awards to exhibitors. Under the

Directory, which brought a burst of spending by *nouveaux riches* tired of Revolutionary austerity, he continued to prosper, and began to gravitate towards the rising star of Bonaparte. (The youngest of the Redouté brothers, Henri-Joseph, served under the general in Egypt as a zoological draughtsman.) Then he went into high gear. Between 1798 and 1804 appeared, or began to appear, since some of the productions came out in instalments over several years, his illustrations for his *Les Liliacées*, for A.P. de Candolle's *Plantes Grasses* (partly undertaken earlier), for André Michaux's *Histoire des Chênes de l'Amérique*, for the same author's *Flora Boreali-Americana*, for Duhamel du Monceau's *Traité des Arbres et Arbustes*, and for Étienne-Pierre Ventenat's *Jardin de la Malmaison*—ultimately a grand total of more than a thousand plates. By the time of the 1805 edition of Rousseau's *Botanique* Pierre-Joseph Redouté was a celebrity, *le Raphaël des fleurs*, and a well-to-do business man with a fashionable clientele, a private apartment in the Louvre, a country residence near Paris at Fleury-sous-Meudon (where Jean-Jacques had once botanized), and a salary of 18,000 francs a year as Joséphine's decorator and flower painter at Malmaison.

Luck and knack for time-serving can partly explain this success. So can hard work and a personality that seems to have been all the more attractive because of a lack of physical grace. In a portrait painted by François Gérard and engraved by C.S. Pradier (the sculptor James Pradier's brother), Redouté is merely rather rugged and unkempt; but if we can trust a description by another contemporary, the memoir-writer Joseph-François Grille, the reality was quite different:

> A dumpy body, limbs like an elephant's, a head as heavy and flat as a Dutch cheese, thick lips, a hollow voice, crooked fingers, a repulsive look, and beneath the skin an extremely delicate sense of tact, exquisite taste, a deep feeling for art, a fine sensibility, nobility of character, and the perseverance needed for the development of genius: such was Redouté, who had all the pretty women in Paris as his pupils.

To this catalogue of contrasts can be added the little visible fact that the apparently uncouth artist had a talent that was both admirable in itself and admirably adjusted to the requirements of his subject matter and of the prevailing period style. His handling of foreshortening and tone values gave his depicted plants a three-dimensional presence that took them well out of the category of the purely decorative or the fuzzily sentimental. Within the limits of his relatively minor genre he was as didactic as the Neoclassical painters of history, although instead of preaching the virtues of ancient Rome he stressed, by a clear portrayal of parts and structures, the merits of the Linnaean system of classification and nomenclature. At the same time he managed—and this was certainly one of the secrets of his popularity—to give his little pictures, executed

COROLLAS *a* Yellow Water-Lily *b* Orange Lily *c* Single Pink *d* Double Pink

a
b
c
d

on vellum in pure, delicately varied water-colour with touches of gouache, an elegance that went well with Directoire and early Empire dress and furniture.

Luck, personality, talent, and hard work would not, however, have produced the reputation and money they did if they had not been helped by an awareness of the importance of moving from the vellum paintings on into the production of plates for books.

Mixed media and master craftsmen

Redouté had developed an interest in etching and related processes as a young man, and shortly after his arrival in Paris in 1782 he had received some instruction in colour-printing from the engraver Gilles Antoine Demarteau, a specialist in aquatint and the crayon manner, a recently invented method of reproducing the texture of chalk drawings. In London a few years later he had discovered another tone process, stipple engraving, popularized by an Italian virtuoso, Francesco Bartolozzi, and by William Ryland (who, overly confident in his skill, had turned to bank-note forgery and had ended his artistic career prematurely on the gallows). In 1796, according to his later claim in a controversy that reached the courts, he had perfected 'a method of our own' of making prints that resembled paintings, 'as can be seen in our *Plantes Grasses, Liliacées*, and other works'. Possibly, although he won his case, all he had actually done was to improve existing practice. But in any event it was true that the illustrations for his major albums, the Rousseau *Botanique* included, were his original water-colours reproduced by a remarkably accurate combination of stipple engraving and single-plate printing.

As in ordinary etching, a sheet of copper was first coated with a mixture of waxes, gums, and resins in order to create an acid resist, or 'ground'. Using a steel needle that cut through the ground and exposed the bright metal, the etcher sketched, often in broken lines, the contours and certain details of the plant. Then, using a small toothed wheel or two needles bound together, he pierced the ground with dots, massing them as required to build up the forms and tones of petals, leaves, and stems. After a bath in a dilute acid that bit into the unprotected copper, the plate was cleaned and subjected to more jagging and flicking by means of a dry-point tool or a special, curved stipple graver. Thus the entire plant was finally represented by a myriad of pits, each a tiny ink-well. The printer applied the colours to their appropriate areas, employing a cloth dabber called a *poupée* (because of a vague resemblance to a rag doll), and he carefully re-inked the plate for each impression. The resulting prints, which were practically monotypes, imitated the subtle gradations of the original

PURPLE FOXGLOVE *Digitalis purpurea* (see page 64)

water-colour; and sometimes Redouté almost converted them into replicas by retouching them with a brush.

The role of the engraver in the process was important, and on the plates for the Rousseau book this is acknowledged by the presence of the names of a number of Parisian craftsmen—Bouquet, Prot, Chailly, De Gouy, Marchand, Masson, Souet, Tassaert—along with that of a craftswoman, Mlle Delelo. But the role of the printer was also considerable, since he had to handle his *poupée* with painterly skill, and this is recognized in the legends by the repetition of *de l'imprimerie de Langlois* or, more explicitly, *imprimée en couleur par Langlois*. Exactly which Langlois he was is a question, for the records of the period list a Langlois *père* as an ink-maker in the rue Saint-Séverin, a Langlois *fils* as a printer and bookseller in the rue Saint-Jacques, and a Hyacinthe-Eloy Langlois as a bookseller at 6 rue de Seine. The last named, however, in spite of his not being described as a printer, is the most interesting, for shortly after the appearance of the 1805 *Botanique* Redouté moved with his wife and three children to a large studio and apartment, admired by a contemporary as 'a temple', at 6 rue de Seine. Moreover, the bookseller Garnery, one of the co-publishers of the album, was the owner of the building. (It still exists in part.)

In 1805 Redouté still had half of his career ahead of him, and he went on being enterprising. He had already done his own publishing occasionally, and one gets the impression that he and possibly Langlois kept control of the sixty-five plates for the *Botanique*, like a modern book-designer with a 'package', and disposed of them at will. After the first two editions, published by Garnery in the rue de Seine and the dealer Delachaussée at 40 rue du Temple, there was the third, published by Baudouin Frères at 36 rue de Vaugirard (this building also still exists); and during the following four years the plates were used for three more printings. In the meantime, in 1817, the artist had begun to bring out his most famous work, *Les Roses*. But he had never been a saver for a rainy day, and so the last part of his life was a long struggle against bankruptcy. He died in the summer of 1840 at the age of eighty-one after having a stroke while —in exact compliance with Rousseau's first instruction to Mme Delessert—he was examining the corolla of a lily.

Roy McMullen

ELDER *Sambucus nigra* (see page 76)

THE FIRST LETTER

22 August 1771

Your notion of diverting your daughter's lively mind and of teaching her to observe such agreeable and varied objects as plants seems to me excellent; I would not have dared to suggest it to you for fear of aping Monsieur Josse.[1] Since the suggestion comes from you, I approve it with all my heart and will offer all the help in my power, for I am convinced that at any age the study of nature abates the taste for frivolous amusements, subdues the tumults of passion, and bestows upon the mind a salutary nourishment by filling it with a subject most worthy of its contemplation.

You have started by teaching your young daughter the names of all the common plants you find about you; this was the very thing you should have done. These few plants which she knows by sight are so many points of comparison for her to extend her knowledge; but they are not sufficient. You ask me for a brief list of the most familiar plants, with the characteristics by which they may be known. I find some difficulty in doing this for you. How can I describe these signs or characteristics clearly yet succinctly in writing? It seems to me impossible unless I employ the appropriate terms; and these terms form a vocabulary apart, which you would not be able to comprehend were it not first explained to you.

Besides, simply to recognize plants and to know nothing but their names cannot but be an excessively dull exercise for such minds as yours, and it is unlikely that it would entertain your

LEAVES *a* Lady's Mantle *b Robinia caragna* *c* Water Crowfoot *d Adiantum reniforme*

a
b
c
d

daughter for long. I suggest that we take a few preliminary notions of plant structure or formation, so that, even should you take but a few steps into the most beautiful, the richest of nature's three kingdoms, you may at least proceed with some knowledge. We need not yet concern ourselves with the nomenclature, which is merely the dry terminology of the botanist. I have always believed that one could be a very great botanist without knowing the name of a single plant; and without wishing to make of your daughter a very great botanist I think, nevertheless, that she will always find it useful to have learned how really to see what she is looking at. Do not, however, be alarmed at the undertaking. You will soon come to realize that it is not a great one. In what I am about to propose to you, you will not find anything complicated or hard to follow; nothing is required but the patience to begin at the beginning; after that you need advance only as far as you wish.

We approach the end of autumn and those plants which are the most simple in their structure are already past. I would, moreover, appreciate a little time to impart some order to your observations. But whilst we await the coming of spring to enable us to begin to follow the course of nature, I shall at least give you some few words of the vocabulary to memorize.

A perfect plant is made up of the root, the stem, branches, leaves, flowers and fruits (for in botany we call the final product of fertilization in both plants and trees the fruit). You already know all that, at least sufficiently to understand the word; but there is an important area which demands closer examination: this is *fructification*, which comprises both the *flower* and the *fruit*. Let us start with the flower, which is the first to appear. It is here that nature has enclosed the culmination of her efforts; it is through the flower that she perpetuates her work; and the flower is also commonly the most brilliant part of the plant, and always the least subject to variation.

PEAR FRUIT *Pyrus communis* (see page 106)

Take a Lily. I think that you will with no difficulty still find one in full bloom. Before it opens out you see at the tip of the stem an oblong greenish bud which becomes white as it is about to open; and when it has completely opened you see that its white outer parts take the shape of a vase divided into several segments. These coloured outer parts, which are white in the Lily, are called the *corolla*, and not, as they are commonly known, the flower, for the flower is made up of several parts of which the corolla is only the foremost.

As you can easily observe, the corolla of the Lily is not of one piece. When it fades and drops, it falls in six separate pieces, which are called the petals. So the corolla of the Lily is composed of six petals. The corollas of all flowers that are thus made up of several pieces are called *polypetalous*. If the corolla should consist of one piece only as, for example, in the Convolvulus, known as the bindweed, it would be called *monopetalous*. Let us return to our Lily.

In the corolla, exactly in the centre, you will find a sort of little column, attached at the base and pointing straight upwards. This column in its entirety is called the *pistil*; it is divided into three parts. 1st, its swollen cylindrical base with three rounded angles, called the *ovary*. 2nd, a thread which rises from the ovary, called the *style*. 3rd, the style is crowned with a sort of capital with three indentations, called the *stigma*. These are the three components of the pistil.

Between the pistil and the corolla you find six more quite distinct parts, which are called the *stamens*. Each stamen is composed of two parts: a thinner one, by which the stamen is anchored to the base of the corolla and which is called the *filament*; and a thicker one at the top of the filament, which is called the *anther*. Each anther is a case which, when it ripens, opens to release a strongly perfumed yellow dust, of which we shall speak later. This dust so far possesses no

ORANGE LILY *Lilium bulbiferum*

name in French; the botanists call it *pollen*, a word which signifies dust.

Such is the general analysis of the parts of the flower. As the corolla fades and falls, the seed head swells to become an elongated triangular capsule, inside which are contained the flat seeds, partitioned into three cells. This capsule which protects the seed is called the *pericarp*. But I shall not undertake the analysis of the fruit yet: that will be the subject of another letter.

The parts I have just named are also to be found in the flowers of most other plants, but they vary in proportion, position and number. By the analogy of these parts and their different combinations the diverse families of the plant kingdom are determined: and these analogies are connected with others in those parts of the plant which seem to have no relation to them. For example, it is a characteristic of all the liliaceous family that there are six petals or divisions of the corolla, and a triangular ovary with three cells; and throughout this most numerous family the roots are always *bulbs*, more or less strongly marked and varied in form or composition. The bulb of the Lily is composed of overlapping scales; in the Asphodel it is a cluster of elongated cloves; in the Saffron there are two bulbs, one upon the other; in the Colchicum they are side by side, but still bulbs.

The Lily, which I have chosen because it is in season and also because of the size of its flower and of its parts, which makes them easier to observe, lacks however one of the essential characteristics of a complete flower: the calyx. The *calyx* is the green part commonly divided into five leaflets, which supports and surrounds the base of the corolla, and which covers it completely before it blooms, as you will have been able to see in the Rose. The calyx, which is a part of almost all other flowers, is absent in most members of the Lily family, such as the Tulip, the Hyacinth, the Narcissus, the Tuberose, etc., and even the Onion, the Leek and the Garlic,

DETAILS OF MADONNA LILY *Lilium candidum* *a* Stamen *b* Petal *c* The six stamens and pistil in normal arrangement, but stripped of corolla *d* The pistil, cut to show hollow centre *e* The pistil *f* Tip of filament and anther, magnified *g* The capsule, cut open *h* Seeds *i* Complete capsule

a
b
c
d
e
f
g
h
i

which are also true members of the Lily family, although they appear so different on first acquaintance. You will also notice that in the whole family the stems are simple, with few branches, the leaves are entire and never indented: observations which confirm analogies between flower and fruit in this family through analogies with other parts of the plant.

If you follow these details with some attention and make yourself familiar with them by frequent observation, you are already in a position to determine by the careful and continued examination of a plant whether it is or is not of the Lily family, even though you do not know the name of the plant. You see that it is no longer a simple labour of the memory, but a study of observations and facts truly worthy of a naturalist. You will not begin by telling all this to your daughter, nor even tell her later when you have been initiated into the mysteries of the plant world; but you will unveil to her by degrees no more than is suitable to her age and sex, leading her to discover things for herself rather than teaching them to her. Farewell, dear cousin. If all this scribble pleases you, I am at your service.

---

[1] M. Josse is a character in Molière's *L'Amour Médecin*, who advises his client to buy some jewellery because he himself is a jeweller.

NARCISSUS *Narcissus tazetta* (see page 32)

THE SECOND LETTER

18 October 1771

Since you comprehend so well, dear cousin, the basic features of plants, even though so faintly defined, that your sharp eye is already able to distinguish a family likeness within the Lily family, and since our dear young botanist takes pleasure in corollas and petals, I am about to suggest another family on which she will henceforth be able to exercise her slender knowledge. I admit the task will be somewhat more difficult, for the flowers are smaller and the foliage more varied, but it will afford both her and yourself equal enjoyment, at least if you have as much delight in following this flowery path as I have in tracing it out for you.

When in gardens the first sunbeams of spring cast their light on the Hyacinths, Tulips, Daffodils, Jonquils and Lilies-of-the-Valley, whose characteristics are already familiar to you, and illuminate your progress, other flowers such as the Wallflowers or Stocks, and Dame's Violet will soon catch your eye and demand closer examination. Should you find double flowers, waste no time in examining them; they are deformed, or, if you prefer, we have embellished them according to our whim: nature is no longer there; she refuses to be reproduced by such deformed monsters; for while the most arresting part, the corolla, is reduplicated, it is at the expense of the more essential organs, which disappear beneath this splendour.

Take, therefore, an ordinary Wallflower, and proceed to examine its flower. First you will perceive an outer part, the calyx, which is

CROCUS (SAFFRON) *Crocus sativa* (see page 32)

lacking in the Lily family. This calyx is of four parts, which we must simply call leaves or leaflets, since we have no proper name for them as we have for the parts of the corolla, the petals. [These 'leaflets' are of course 'sepals'. Ed.] These four parts are usually in unequal pairs, in that there are two smaller leaflets alternating with two larger ones, which are noticeably enlarged at their bases by swellings on their outsides.

Within this calyx, you will perceive a corolla composed of four petals, whose coloration I may ignore as it is not a characteristic. Each of these petals is attached to the receptacle or base of the calyx by a narrow pale organ termed the *claw* and extends beyond the calyx in a broader more highly coloured part called the *lame.*

In the centre of the corolla lies an elongated pistil which is more or less cylindrical and terminates in a very short style which itself is terminated by a *forked*, oblong stigma – which is to say that it is divided into two lobes, each mirroring the other.

If you examine carefully the respective positioning of the parts of the calyx and the parts of the corolla you will observe that each petal, rather than corresponding precisely with the leaflets of the calyx, is on the contrary positioned between two of them so that it corresponds with the gap between them. This alternate positioning occurs in every species of flower where there are an equal number of petals in the corolla and leaflets in the calyx.

Finally we must mention the stamens. You will find them in the Wallflower six in number, as in the Lily family, but here they are not of equal size, nor is each alternate one of unequal size; for you will see only two, opposite each other, distinctly shorter than the other four which separate them, and which are in turn separated in two pairs.

At the moment I shall not enter into the details of their structure and position; but I would draw to your notice that if you look care-

GARLIC *Allium sativum* (see page 32)

fully you will discover for what reason these two stamens are shorter than the others, and why two of the leaflets of the calyx are more rounded, or to use a botanical term more gibbous, while the other two are more flattened.

We must not end the story of our Wallflower with the analysis of its flower, but must wait until the corolla fades and falls, which happens quite soon, and then observe what becomes of the pistil, which is made up, as we have previously said, of the ovary (or pericarp), the style and the stigma. The ovary becomes much elongated and swells a little as the fruit ripens. When this ovary, or fruit, is ripe, it takes on the form of a sort of flat pod, named a *siliqua*.

This siliqua is made up of two cells, placed one above the other, and separated by a very fine skin called the *septum*.

When the fruit is completely ripe, the cells open from bottom to top to liberate the seeds, and their top ends remain attached to the stigma.

Flat, round seeds are then visible lying on the two faces of the septum, and if one carefully examines how they are attached one sees that each seed is held by a short pedicel alternately to the right and left of the seam of the septum, that is on the two edges by which it was, so to speak, sewn to the cells before their separation.

I greatly fear, dear cousin, that I may have wearied you by this long description, but it has been necessary in order to give you the essential characteristics of the large family which, in almost all botanical systems, forms a whole class; and this description, though somewhat hard to follow here without illustration, will, I trust, become more clear to you when you follow it carefully whilst observing the actual subject.

The great number of species which comprise the family of crucifers have caused botanists to subdivide it into two sections which, despite the similarity of the flowers, differ considerably in the fruit.

WALLFLOWER *Cheiranthus cheiri* *a* Petal *b* The flower, stripped of petals *c* The flower, stripped of petals and calyx *d* Stamen *e* The pistil *f* The siliqua

a
b
c
d
e
f

The first section consists of crucifers with *siliquae*, such as the Wallflowers of which I have just been speaking, Dame's Violet, Watercress, Cabbages, Rapes, Turnips, Mustard, etc.

The second section is made up of crucifers with *siliculae*, where the siliqua is small and very short, almost as wide as it is long, and internally partitioned in a different way; examples of this are the Orleans Cress, commonly called *Twisted Nose*; Thlaspi, which gardeners call Penny-cress; Cochlearia [Scurvy-grass]; Lunaria [Honesty], which, although it has a very large pod is nevertheless still a silicula, since its length hardly exceeds its width. If you are acquainted neither with Orleans Cress, nor Cochlearia, nor Thlaspi, nor Lunaria, at least I presume you will know the Shepherd's Purse, which is such a common garden weed. Well, cousin, the Shepherd's Purse is a crucifer with a silicula which is triangular. By this you may form some idea of the rest until such time as they come to hand.

But it is time to let you draw breath, the more so in that, before the season enables you to make use of it, this letter will, I hope, be followed by several others in which I shall be able to add what remains to be said of importance about crucifers, and which I have not mentioned here. But perhaps it is a good moment to warn you, without further delay, that in this family and in many others you will often find flowers far smaller than Wallflowers, and sometimes so small that you will scarcely be able to observe the details of their structure without the help of a lens – an instrument the botanist cannot do without, any more than he can do without a needle, a scalpel and a pair of good sharp scissors for dissection.

In the belief that your maternal zeal will bring you thus far, I picture to myself a charming scene with my beautiful cousin busy with her glass taking apart heaps of flowers a hundred times less flourishing, less fresh and less agreeable than herself.

Farewell, cousin, until the next chapter.

DAME'S VIOLET *Hesperis matronalis*

THE THIRD LETTER

16 May 1772

I am assuming, dear cousin, that you received my last communication, even though you did not mention it in your second letter. In response to the latter, I hope that, as you wrote, your mother[1] is quite recovered and set out in a good state of health for Switzerland, and I trust that you will not fail to inform me of the effects of this journey and of the waters she is going to take. As Tante Julie[2] was to depart with her, I have entrusted to Monsieur G.,[3] who is returning to Val-de-Travers, the little pressed flower collection[4] which is intended for her, and I have addressed it to you, so that you may receive it in her absence and make use of it, should there happen to be among these disordered specimens anything of use to you. I must add that I do not grant you any rights over this trifle. You have rights over him who made it, the strongest and dearest I know; but as for the pressed flowers, they were promised to your sister when she went on botanical walks with me at La Croix de Vague, whilst you were dreaming of nothing less in those walks with Grand'maman[5] in Vaise, where my heart and my feet followed you. I blush for having kept my word so late and so ill, but it is she, when all is said, and not you, who have my word and therefore prior claim in this. As for you, dear cousin, if I do not promise you a collection of pressed flowers made with my own hands it is because I want you to have a more precious one still, from the hands of your daughter, if you continue to pursue with her this gentle and charming study which

LEAVES *a* Elder *b* Lily *c* Woodruff *d* Honeysuckle

a
b
c
d

fills with interesting observations on nature those empty moments devoted by others to idleness, or worse. For the moment, let us once more take up the interrupted theme of our plant families.

My intention first of all is to describe to you six of these families, in order to make you familiar with the general structure of the characteristic parts of plants. You have already made acquaintance with two; there are four remaining that you must have the patience to study: after which, leaving for a while the other branches of this large family tree and passing to the examination of the various organs of fertilization, we shall ensure that, whereas you may not be able to recognize many plants, you will never find yourself in unknown territory among the manifestations of the plant kingdom.

But I warn you that if you study books and learn the standard nomenclature, you will know many names and have little understanding; that which you have will become confused; you will not easily follow my line of thought nor that of others, and at best you will have only a knowledge of words. Dear cousin, I am jealous of being your only mentor in this field. When the time comes, I shall suggest books for you to consult. Until that time, have the patience to read only in the book of nature and to make use only of my letters.

Peas are now in full fruit. Let us seize the moment to study their characteristics, which are among the most curious to be found in the world of botany. As a general rule, all flowers can be grouped into regular or irregular categories. The former are those in which all parts of the flower grow out evenly from the centre, and whose outer extremities would therefore form the circumference of a circle. Their uniformity is such that the eye on observing a flower of this type would distinguish neither top nor bottom, right nor left: the two families we have so far examined are of this type. But you will see at a glance that the flower of the Pea is irregular; that one can easily distinguish in the corolla the longer part, which must be

IBERIS *Iberis sempervirens*

uppermost, from the shorter, which must be lowermost, and that one knows very well when looking at the flower whether one is holding it the right way up or upside down. Thus, in examining an irregular flower, whenever we speak of the top and the bottom, we suppose it to be in its natural position.

As the flowers of this family are of a very particular construction, it is not only essential to have several Pea flowers and to dissect them one by one in order to study all their parts successively; one must also follow the progress of fructification from the first flowering to the ripening of the fruit.

You will find that the calyx is *monophyllic*, that is, formed of a single piece terminating in five very distinct points, of which the two at the top are somewhat broader and the three below somewhat narrower. This calyx is curved downwards, as is the pedicel which bears it; this pedicel is very flexible, very mobile, so that the flower bends easily to the breeze and usually turns its back on wind and rain.

After examining the calyx, one removes it, gently tearing it off so that the rest of the flower remains intact, and then you clearly perceive that the corolla is polypetalous.

Its chief component is a large, broad petal which covers the others and lies uppermost in the corolla; because of this this large petal has been given the name of the *pavilion*. It is also called the *standard*. One would have to close the eyes and mind not to see that this petal acts like an umbrella, guaranteeing those it protects against the major ravages of the climate.

Removing the standard as you did the calyx, you will notice that it is buttressed on either side by a little cushion in the side petals, so that its situation cannot be disturbed by the wind.

This removal of the standard exposes the two side petals to which it was attached by its cushions. These side petals are called the

PEA FLOWER *Pisum sativum*

wings. You will discover, when detaching them, that they are embedded even more firmly in what remains of the flower and cannot be separated from it without some effort. In this way the wings are scarcely less useful in protecting the sides of the flower than is the standard in covering it.

After removing the wings you can see the final component of the corolla; this part covers and protects the centre of the flower, and envelops it, particularly from beneath, as assiduously as the other three petals the top and the sides. This last part, named by reason of its shape the *keel* [Rousseau calls it *nacelle* or *dinghy*. Ed.], is like a strongbox in which nature has deposited her treasure to shield it against the violations of wind and water.

After a close examination of this petal, pull it gently from below, gripping it lightly by the narrow finger-hold it offers you at its base, to avoid removing what it conceals. I am convinced that at the moment when this final petal is compelled to let go and to reveal its mystery, you will not be able to repress a cry of surprise and wonder.

The young fruit which the keel protected is constructed in the following manner. A cylindrical tissue surmounted by ten quite distinct threads surrounds the ovary, that is, the embryonic pod. These ten threads are so many stamens which are united at their base around the ovary and terminate at their top in an equal number of yellow anthers, whose pollen will fertilize the stigma which caps the pistil and which, even though it is coloured yellow by the pollen which clings to it, can easily be distinguished from the stamens by its shape and roundness. And so these ten stamens create yet another and final shield around the ovary to preserve it from external injury.

If you look closely, you will discover that these ten stamens only appear to form a unity at the base; for, in the upper part of this cylinder there is one piece or stamen which while at first seeming to adhere to the others, in due measure, as the flower fades and the

DETAILS OF PEA FLOWER *Pisum sativum* *a* Complete flower *b* The calyx and ovary, stripped of petals and stamens *c* The flower, stripped of calyx lobes *d* The standard, spread out *e* Wing *f* The keel *g* The bundle of stamens, spread out *h* The ovary *i* The pod *k* The pod, split open *l* Seed (pea)

a
b
c
d
e
f
g
h
i
k
l

fruit swells, detaches itself, leaving an opening above through which the swelling fruit may plump itself out, progressively opening out and stretching the cylinder which would otherwise compress and strangle it and prevent it from growing and becoming fruitful. If the flower is not far advanced, you will not see this stamen detached from its cylinder; but pass a needle through the two small holes which you will find near the receptacle at the base of the stamen, and you will soon find the stamen with its anther following the needle and detaching itself from the other nine. These other nine will continue to form a unit until they fade and shrivel, by which time the ripened fruit has become a pod and has no further need of them.

This *pod*, into which the ovary is transformed as it ripens, differs from the *siliqua* of the crucifers in that in the *siliqua* the seeds are attached alternately to the two seams of the septum, whereas in the *pod* they are attached only at one side, that is to say to only one of those seams. It is true that they do grow from both halves of the pod, but always on the same seam. You will understand this distinction perfectly if you open at the same time the *pod* of a Pea and the *siliqua* of a Wallflower, making sure that neither is quite ripe, in order to be certain that on opening the fruit the seeds are still attached by their stalks to their seams and cells.

If I have been able to make myself clear, dear cousin, you will appreciate what amazing precautions nature has taken to bring the embryo of the Pea to maturity and to protect it, particularly when the rain is heaviest, from the effects of humidity, which are death to it, without nevertheless enclosing it in a hard shell which would have made it a different sort of fruit. The supreme maker, attentive to the conservation of all living things, has taken great care to protect the processes of fructification from the dangers that might beset them; but he seems to have been doubly careful of those which

BROOM *Sarothamnus scoparius* (see page 54)

serve to nourish men and animals, as do the majority of the leguminous family. The flowers are called *papilionaceous*, for in shape they seem to bear some resemblance to a butterfly. They normally have a *standard*, two *wings*, and a *keel*, which means that they generally bear four irregular petals. But there are some genera in the family in which the keel is divided lengthwise into two, which almost join below, and these flowers actually have five petals; others, such as the Red Clover have all their petals united, and although they are papilionaceous they are nevertheless monopetalous.

The papilionaceous or leguminous family is one of the most numerous and one of the most useful. Beans, Brooms, Lucerne, Sainfoin, Lentils, Vetches, Tares and others, which characteristically have a spiralled keel that one might at first believe to be accidental. There are also trees, among them the one commonly called the Acacia (which is not the true Acacia), the Indigo Tree and the Liquorice: but we shall speak later in greater detail of all that. Farewell, dear cousin, I embrace all that you hold dear.

---

[1] Julianne-Marie Boy de La Tour.
[2] Mme Delessert's sister.
[3] Frédéric Guyenet.
[4] This collection is preserved today in the central library of Zurich.
[5] Nickname given by Rousseau to Elisabeth, youngest sister of Mme Delessert.

RED CLOVER *Trifolium pratense*

THE FOURTH LETTER

19 June 1772

You have relieved me of my worries, dear cousin; but I still remain perturbed by these stomach pains, these abdominal twinges, which your mama feels recurring when she sits down to write. If it is only a bilious attack, the journey and the waters will suffice to rid her of it; but I much fear that there may be in these attacks some local cause which will not be so easily overcome, and which will always demand great prudence on her part, even after she is cured. I await your news of this journey, as soon as you have any; but I am insistent that your mama write to me only to tell me of her complete recovery.

I cannot understand why you have not yet received the collection of pressed flowers. Believing that Tante Julie had already left, I entrusted the package to Monsieur G. to send to you when he passed through Dijon. As far as I can ascertain it has come neither into your hands nor into those of your sister, and I cannot imagine what can have become of it.

Let us talk of plants, whilst the season for observing them invites us. Your answer to the question I posed on the stamens of the crucifers is completely right, and proves to me that you have understood me, or rather that you have listened to me; for you need only listen to understand. You have accounted very well for the swollen appearance of two of the leaflets of the calyx and for the relative shortness of two of the stamens of the Wallflower, resulting from the curve of these two stamens: however, one more step would have brought you to the real cause of this formation: for, if you seek further to

FALSE ACACIA *Robinia pseudoacacia* (see page 54)

discover why these two stamens are thus curved, and in consequence shortened, you will find a little gland [the nectary. Ed.] situated on the receptacle between the stamen and the seed; and it is this gland which, by forcing the stamen away and compelling it to curve, is responsible for shortening it. On the same receptacle there are two more glands, one at the base of each of the pairs of large stamens; but, as they do not force them into a curve, so they do not shorten them, because unlike the first two, these glands lie not inside, between the stamen and the seed, but outside, between the pair of stamens and the calyx. And so these four stamens point upwards and grow straight, outstripping those which curve downwards, and thus appear the longer because they are the straighter. These four glands, or at any rate traces of them, can be found more or less distinctly in nearly every cruciferous flower, and in some even more distinctly than in the Wallflower. If you still wonder what these glands are for, I will answer that they are one of the mechanisms destined by nature to unite the vegetable and animal kingdoms, and allow them to intermingle. [Rousseau refers here to the fact that insects feed on the nectar. Ed.] But we anticipate. Let us return for the present to our families.

The flowers which I have so far described to you have all been polypetalous. Perhaps I ought to have started with the regular monopetalous flowers, whose structure is simpler by far: this very simplicity is what made me hold back. The regular monopetals form less a family than a great nation among which one can enumerate several quite distinct families; to such a degree that, were one to include them all under a common heading, one would have to use such general and vague terms that one would appear to be saying something when in reality one would be saying almost nothing. It is better to confine oneself within narrower limits, which can be defined with greater precision.

LUCERNE *Medicago sativa* (see page 54)

Among irregular monopetals, there is one family whose appearance is so marked that one can easily recognize its members on sight. This is the family commonly called mouth-shaped flowers, for their flowers are split into two lips which, whether they open naturally or when lightly squeezed, create the impression of a gaping mouth. This family is subdivided into two sections or families: one, *labiate* flowers, or lipped flowers; the other, *personate* flowers, or masked flowers: for the Latin word *persona* signifies a mask, a name which certainly fits the majority of people amongst us, whom we call *persons*. What is common to the whole family is not only the possession of the monopetalous corolla, which, as I have already told you, is divided into two lips or jaws, the upper named the *helmet*, and the lower the *lip*; but also the four stamens which are positioned almost in one row, separated into two distinct pairs, one pair longer and the other shorter. Careful study of the flower itself will explain these characteristics better than any description can.

Let us first examine the *labiates*. I would happily suggest as an example the Sage, which we can find in practically every garden. But the peculiar and bizarre construction of the stamens, which has caused some botanists to exclude it from the labiate family, despite the fact that nature has seemingly placed it there, obliges me to seek another example in the Dead-Nettles, and particularly in the species commonly called the *White Dead-Nettle*, which botanists prefer to call the *White Lamium*, since in reproductive terms it has nothing in common with the Nettle, notwithstanding the similarity of its foliage. The White Dead-Nettle, which is common everywhere and has a very long flowering season, should not be difficult for you to find. Without pausing to describe the elegant arrangement of the flowers, I shall confine myself to their structure. The White Dead-Nettle possesses a labiate monopetalous flower, with a concave helmet, curved like a vault to protect the rest of the flower,

WHITE DEAD-NETTLE *Lamium album* *a* Complete flower *b* The calyx *c* The corolla *d* Stamen *e* The pistil *f* The ovaries

a
b
c
d
e
f

especially the stamens, which are all four quite tightly squeezed under the shelter of its roof. You will easily discern the longer and the shorter pair, and in the middle of the four the similarly coloured style, which is distinguished by the fact that its tip is simply forked, instead of bearing an anther as do the stamens. The lip, meaning the lower lip, is doubled back and hangs down, and this arrangement allows a glimpse almost into the depths of the corolla. In *Lamiums* this lip is divided lengthwise down its centre; but such is not the case with the other labiates.

If you tear off the corolla, you will tear off with it the stamens which are attached to it by their filaments and not to the receptacle, to which only the style will remain attached. On examining how the stamens are attached in other flowers, you will discover that they are usually attached to the corolla in monopetalous flowers, and to the receptacle or to the calyx in a polypetalous corolla; so that, in the latter case, it is possible to tear off the petals without tearing off the stamens. From this observation we can form an elegant, easy and even fairly certain rule for knowing whether a corolla is of one piece or of several, even when it is difficult to be immediately sure.

When the corolla has been removed there is a hole at its base, for it was attached to the receptacle, leaving a circular aperture through which the pistil and its surrounding parts penetrated the tube and the corolla. What surrounds the pistil in Lamiums and in all labiates is four embryos which will become four naked seeds, that is, with no covering: so that when these seeds ripen, they detach themselves and fall separately. These are the characteristics of labiates.

The other family or section, that of the *personates*, differs from the labiates firstly in the corolla, whose two lips are not normally open or gaping, but are closed and joined, as you can see in the garden flower known as *Antirrhinum* or *Snapdragon*, or failing that in Toadflax, that spurred yellow flower so common at this time of the

SNAPDRAGON *Antirrhinum majus*

year in the country. But a more accurate and sure point of identification lies in the fact that instead of having four naked seeds at the base of the calyx as in the labiates, the personates all possess a capsule which encloses the seeds and opens only when they are ripe so that they can be scattered. I would add to these characteristics that a large number of labiates are either sweet-smelling and aromatic plants, such as Oregano, Marjoram, Thyme, Wild Thyme, Basil, Mint, Hyssop, Lavender, etc.; or they are foul-smelling and malodorous plants, such as several species of Dead-Nettle, Woundworts, Toadstones, Horehound: only a very few, such as the Bugle, Self-heal, Scullcap have no scent at all; the personates, on the other hand, are mostly odour-less plants, such as the Snapdragon, Toadflax, Eyebright, Lousewort, Crested Cow-Wheat, Broomrape, Ivy-leaved Toadflax, Purple Toadflax, Foxglove: within this family I scarcely know any with an odour, save the Figwort, which smells foul and is not aromatic. I can hardly mention here any plants but those which are probably unknown to you, but which gradually you will get to know, and whose family, at least, you will be able to determine when you encounter them. I would even like you to try to determine the family or its subdivision by the features, and to practise judging, at a glance, whether the mouth-shaped flower you perceived is a labiate or a personate. The external appearance of the corolla may suffice to guide your choice and you will then be able to verify it by removing the corolla and looking into the base of the calyx; for if you are correct, the flower which you have identified as labiate will expose four naked seeds, and that which you identified as personate will display a pericarp: the opposite would prove that you were mistaken, and by a second examination of the same plant, you will guard against a similar error in the future. There, dear cousin, is something to occupy you for a few strolls. I shall not long delay before preparing you for the walks to come.

DETAILS OF SNAPDRAGON *Antirrhinum majus* (see page 62) *a* Complete flower *b* The calyx and style *c* The corolla, seen from behind *d* The corolla, split open *e* Stamen *f* The pistil *g* The capsule *h* The capsule, cut open *i* The capsule, cut open along its length

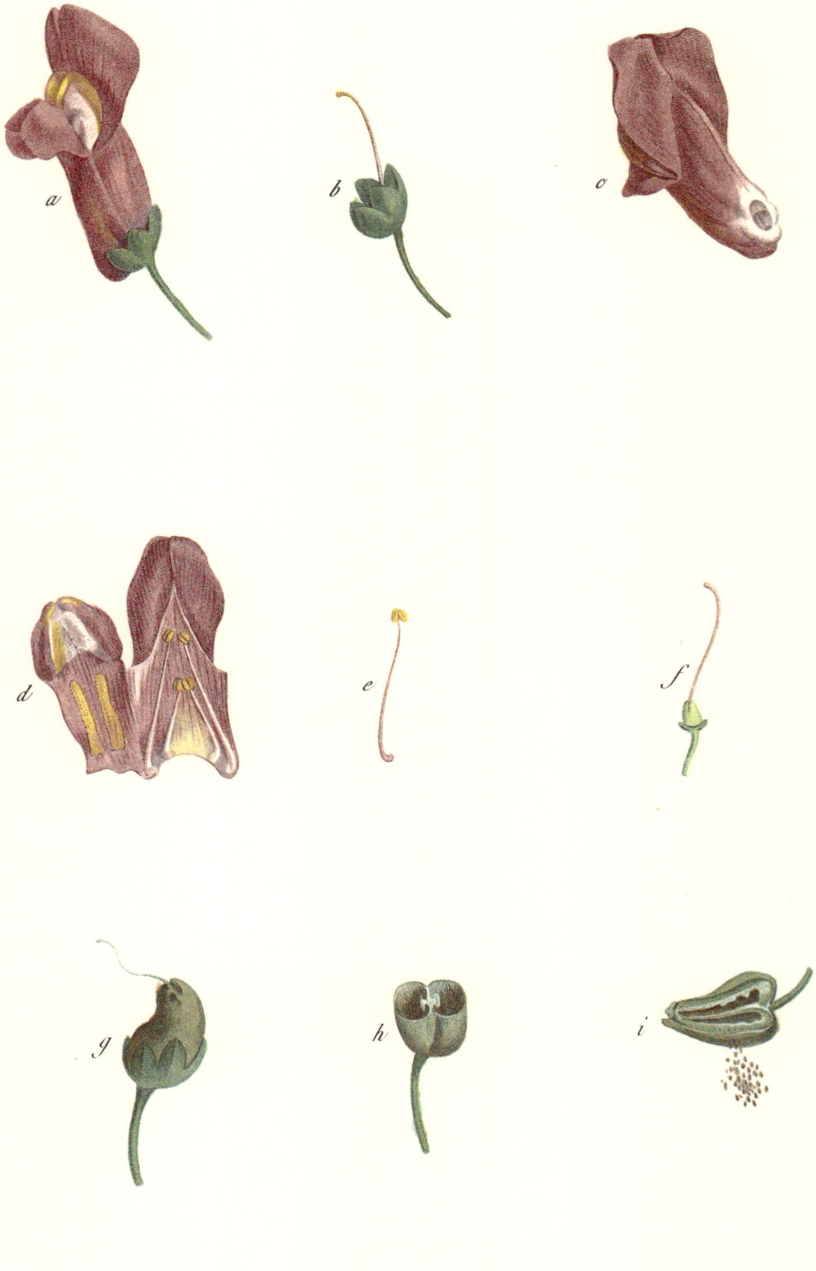
a
b
c
d
e
f
g
h
i

THE FIFTH LETTER

16 July 1772

Thank you, dear cousin, for the good news you gave me of your mama. I had hoped for the beneficial effect of a change of air and I expect no less from the waters and above all from the strict regime prescribed during their taking. I am touched to be remembered by this kind friend and I beg you to thank her for me. But I most definitely do not wish her to write to me during her stay in Switzerland and if she wishes to send me news of herself directly she has with her a good secretary[1] who will perform the task admirably. I am delighted, rather than surprised, that your sister is such a success in Switzerland: apart from the charms natural to her age and her warm vivacious gaiety, she has in her nature a fundamental gentleness and evenness of temper, surely learned from you, and which I have sometimes seen her display so charmingly to Grand'-maman. If she settles in Switzerland you will both lose a great sweetness in your lives, and she above all will lose benefits which it will be hard to replace.

But your poor mama, who found separation from you, even when you were only next door, so cruel, how will she bear being so far from your sister? It is from you that she will draw her strength and her comfort. You are preparing a most precious one for her as you mould in your gentle hands the good sound material of your most cherished daughter who, I have no doubt, will become under your care as full of good qualities as of charm. Ah, cousin, what a happy mother is yours!

TOADFLAX *Cymbalaria muralis* (see page 62)

Do you know, I am becoming anxious about the little collection of pressed flowers. I have heard nothing of it whatsoever, although I did have news of Monsieur G. after his return, from his wife who, for her part, did not say a word about the collection. I asked her for news of it; I am waiting for her answer. I am very much afraid that, not passing through Lyons, he may have given the parcel to someone or other who, knowing it to be dried grasses, took it for a bundle of hay. But if, as I still hope, it eventually reaches your sister Julie or you, you will find that I have taken some care over it. It is a loss which, though small, I should not easily be able to replace quickly, above all because of the catalogue on which I wrote various little explanations as I went along, and of which I have no copy.

Do not be upset, dear cousin, if you did not see the glands of the crucifers. Great and keen-eyed botanists have been no better at noticing them. Tournefort[2] himself mentions them not at all. They are only distinct in a few genera, although traces of them are found in almost all; and by repeatedly analysing cruciform flowers, and always noticing the irregularities of the receptacle, it was found on examining them individually that these glands were present in the majority of genera, and by analogy one may also assume that they are present also in those where one cannot distinguish them.

I realize that it is frustrating to take so much trouble and not learn the names of the plants which one is studying; but I frankly confess to you that it is no part of my plans to spare you this small vexation. It has been said that botany is nothing but a science of words, which merely exercises the memory and teaches only the naming of plants. For my part I know of no worthwhile study which is only a science of words; and to whom, pray, shall I award the name of botanist: to him who can let fall a name or a phrase on seeing a plant, without knowing anything about its structure, or to him who, while being thoroughly conversant with its structure, is nonetheless ignorant of

SAGE *Salvia officinalis* (see page 60)

the quite arbitrary name which is bestowed upon the plant in this country or that? If we are only giving your children an amusing occupation, we are failing to achieve the more important part of our intention, which is, whilst granting them entertainment, to exercise their intelligence and accustom them to careful observation. Before teaching them to give a name to what they see, let us start by teaching them to see. This science, which has been forgotten in every system of education, must form the most important principle of theirs. I can never reiterate this enough; teach them never to be satisfied with words, and to believe that they know nothing whilst it resides only in their memory.

However, in order not to be too unkind, I shall nevertheless give you the names of some plants whereby, if they are pointed out to you, you can easily verify my descriptions. You did not, I imagine, have a White Dead-Nettle before your eyes when you read my analysis of the labiates; but you had only to send to the herbalist at the corner for some freshly picked White Dead-Nettles and to apply my description to the flower, after which, by examining the other parts of the plant in the manner in which we are about to embark, you would know the White Dead-Nettle infinitely better than the herbalist who supplies it will know it in his lifetime; and shortly we shall find the means of doing without the herbalist: but first we must complete the study of our families. And so I come to the fifth which, at this very moment, is in full bloom.

Imagine a long and fairly straight stem, with alternate leaves, normally rather finely indented, enclosing, at their bases, stalks which spring from the axils. From the top of this stalk radiate, as if from a centre, several pedicels or spokes, which spread out in a regular, circular fashion like the ribs of a parasol, and crown this stalk in the shape of a more or less wide mouthed vase. On some occasions, these spokes have an unfilled space in their midst, and

FIELD ERYNGO *Eryngium campestre* (see page 78)

they then more closely resemble the hollow of a vase; sometimes this space in the centre is furnished with shorter spokes, growing at less of an angle, which adorn the vase and, together with the outer set, create a form which approximates to the shape of a hemisphere with the convex side uppermost.

Each of these spokes or pedicels terminates at its tip not as yet in a flower, but in another arrangement of smaller spokes, crowning the first in the very same manner as the first crowned the stalk.

So here we find two similar and successive arrangements: the one of large spokes terminating the stalk, and the other of smaller but similar spokes terminating each of the larger ones. The spokes of the little parasols are not further divided, but each of them forms the pedicel of a small flower of which we shall speak presently.

If you are able to form a picture of what I have just described to you, you will have in your mind an idea of the way in which the flowers are organized in the *umbelliferous* family (literally 'parasol-carrying' family: for the Latin word *umbella* means parasol).

Although this regular arrangement of the inflorescence is a striking one and more or less constant among all the umbellifers, it is, however, not this which constitutes the essential character of the family. This character is drawn from the actual structure of the flower, which I must now describe for you.

But for greater clarity it is necessary for me to categorize for you all plants in a general way according to the relative positions of the flower and the fruit; a categorization which greatly facilitates their methodical classification, whichever system you may choose.

In the great majority of plants, for example the Pink, the ovary is quite clearly confined within the corolla. We shall call these *inferior flowers*, because the petals enclosing the ovary grow from a point beneath it. [N.B. In modern terminology the organization of these plants is known as ovary superior. Ed.]

HEMLOCK *Conium maculatum* (see page 80)

In other plants, of which there are quite a large number, the ovary is found not within the petals but below them; as you can see in the Rose: for the hip, which is its fruit, is that green and swollen part that you perceive below the calyx, and therefore also below the corolla, which thus surmounts this ovary and does not envelop it. I shall call these flowers *superior* [ovary inferior. Ed.], because the corolla lies above the fruit. One could invent more Gallicisms; but it seems to me preferable to remain as close as possible to the terms used in botany, so that, without needing to learn either Latin or Greek, you may reasonably well understand the vocabulary of this science, culled pedantically from these two languages, as if, to understand plants, one needed first to be a learned grammarian.

Tournefort made the same categorization, though in other terms: in the case of the *inferior flower*, he said that the pistil became the fruit; in the case of the *superior flower*, he said that the calyx became the fruit. This mode of expression may have been as clear, but it was certainly not so accurate. Be that as it may, here is an opportunity for your young pupils to practise, when the time comes, how to disentangle similar ideas, presented in quite different terms.

I shall now disclose to you that umbellifers have a *superior flower*, or one positioned above the fruit. The corolla of this flower has five so-called regular petals, although on those flowers which lie at the edge of the umbel the two petals which are on the outer edge are frequently larger than the other three.

The formation of these petals varies according to the genus, but it is most commonly heart-shaped; the claw, which grows over the ovary, is very narrow; the petal broadens out; its edge is *emarginate* (shallowly scalloped), or else it terminates in a point which, turning back on itself, once again imparts an emarginate appearance to the petal, although were it to be straightened out one would see that it was pointed.

FLOWER STRUCTURES *a* Catkin (male): Walnut *b* Raceme: Black-currant *c* Panicle: Lilac *d* Composite head: Cupidone *e* Spike: Wheat

a

b

c

d

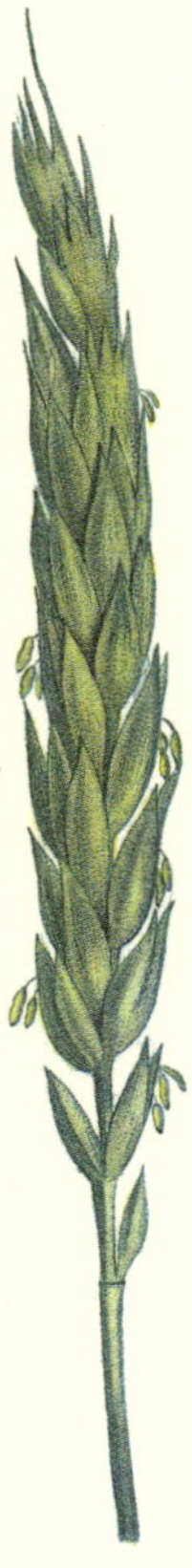

e

Between each petal lies a stamen whose anther protrudes beyond the corolla, making the five stamens more visible than the five petals. I am not making any reference to the calyx here, for it is not distinct among the umbellifers.

From the centre of the flower arise two styles, which are quite prominent and are each furnished with a stigma; after the petals and stamens fall, these styles remain to crown the fruit.

The most usual shape for this fruit is a somewhat elongated oval, which at maturity splits in half and divides into two naked seeds attached to the pedicel, which by some marvellous process divides in two just like the fruit, and holds the seeds apart until they fall.

These proportions all vary according to the genus; but that is the most common arrangement. One must, I admit, possess a most attentive eye to distinguish such small objects without the aid of a magnifying glass; but they are so worthy of attention that one does not begrudge the effort.

These are the special characteristics of the umbelliferous family: a superior corolla with five petals, five stamens, two styles growing from a naked *dispermous* fruit, that is one *composed of two seeds* adjoining each other.

Every time that you find these characteristics combined in one inflorescence, be assured that the plant is an umbellifer, even though the structure of the plant has nothing in common with that which we have described above. And even should you find that this arrangement of parasols conforms to my description, you may be sure that you are deceived if it is contradicted by a closer examination of the flower.

If it should happen, for example, that on a walk just after reading my letter you should come across an Elder still in flower, I am almost certain that at a first glance you would say, here is an umbellifer. On looking at it you would find a large umbel, a small umbel,

GOLDEN CHERVIL *Chaerophyllum auratum* (see page 80)

small white flowers, a superior corolla and five stamens; it is most assuredly an umbellifer. But let us look again: I pluck a flower.

First, instead of five petals, I find a corolla, with five divisions it is true, but nevertheless all of one piece. Now, the flowers of umbellifers are not monopetalous. There are indeed five stamens, but I see no styles; and I see three stigmas more often than two, three seeds more often than two. Now, umbellifers never have more nor less than two stigmas, nor do they ever have more nor less than two seeds for each flower. Finally, the fruit of the Elder is a soft berry, and that of the umbellifers is dry and naked. The Elder is thus not an umbellifer.

If you now retrace your steps and examine the arrangement of the flowers more carefully, you will see that it is only in appearance that of the umbellifers. The larger spokes, instead of radiating perfectly from the same centre, take their point of departure some higher up, some lower down; the small ones are even less regular: in none of this do they follow the invariable structure of the umbellifers. The flowers of the Elder are arranged in a *corymb* or cluster, rather than in an umbel. Thus by making a blunder sometimes, we learn to see with more accuracy.

The Eryngium, on the contrary, scarcely resembles an umbellifer; and yet it is one, for it bears all the characteristics in its inflorescence. Where can I find the Eryngium? you will say. Anywhere in the countryside. All the highways are carpeted with it on both sides: any countryman can point it out to you, and you would almost recognize it yourself by the bluish or sea-green colour of its leaves, with their stiff prickles, and by their texture, smooth and leathery like parchment. But one can ignore such an intractable plant; it has not beauty enough to compensate for the injuries one inflicts on oneself whilst examining it; and were it one hundred times as pretty, my little cousin, with her small sensitive fingers,

LEAVES *a* False Acacia *b* Pear *c* Orange *d* Clover *e* Trefoil

a
b
c
d
e

would soon be disheartened if she were to handle a plant of such vile temper.

The umbellifer family is numerous, and so natural that it is very hard to distinguish the genera: they are brothers whose great likeness often causes them to be confused with one another. To help in recognizing them, some broad distinctions have been drawn, which are sometimes useful, but on which one must not depend too much either. The focal point from which the spokes of both the large and the small umbel radiate is not always naked; it is sometimes encircled by leaflets, as by a ruff. These leaflets have been given the name of *involucre* (envelope). When the large umbel sports a ruff, this ruff is called a *large involucre*: *little involucres* are those which sometimes encircle the small umbels. This gives rise to three groups of umbellifers:

1 Those with a large involucre and little involucres. 2 Those with only little involucres. 3 Those with neither large nor small involucres.

It would appear that there ought to exist a fourth group with a large involucre and no small ones; but no genus is known in which this is consistently the case.

Your remarkable progress, dear cousin, and your patience have so emboldened me that, with no regard for your suffering, I have dared to describe the family of the umbellifers without letting you set eyes on a single example; and this must needs have made much greater demands on your concentration. I am certain, however, that reading as you do, after you have looked over my letter once or twice, you will not fail to recognize an umbellifer in flower when you come across one; and, at this time of the year, you cannot fail to find several in gardens and in the countryside.

They bear, for the most part, white flowers. Such are the Carrot, Chervil, Parsley, Hemlock, Angelica, Cow-Parsley, Water-Parsnip, Boucage, Rough Chervil, Rock Samphire, etc.

FRUITS AND SEEDS *a* Drupe, cut open: Plum *b* Pome, cut open: Apple *c* Berry, complete, and cut open: Grape *d* Pod, split open: Pea *e* Pod, closed: Pea *f*, *g* Siliqua, split, and closed: Rockett *h* Berry: Buckthorn *i* Multi-celled capsule: Hibiscus *k* Capsule: Pink *l* Naked seeds: labiate

a
b
c
d
e
f
g
h
i
k
l

A few, such as Fennel, Dill, Parsnip, have yellow flowers; a few have flowers with a reddish tinge, and there are none of any other colour.

This, you will tell me, gives a fine general idea of the umbellifers; but how will all this vague information ensure that I do not confuse Hemlock with Chervil and Parsley which you have just mentioned together? The most humble cook will know more about this than we with all our learning. You are right. But nevertheless, if we start out with detailed observations, soon, swamped by numbers, our memory will desert us, and we shall lose our way with our first few steps into this vast domain; whereas if we begin by clearly discerning the main highways we shall rarely stray into byways, and we shall find our way everywhere without much difficulty. However, let us make an exception in view of the usefulness of the subject, and not expose ourselves, as we explore the vegetable kingdom, to eating a Hemlock omelette through lack of knowledge.

The small garden Hemlock is an umbellifer, just like Parsley and Chervil. Like these other two, it bears white flowers [The Parsley flower has a yellowish tinge. But several umbellifers appear yellow because of the ovary and anthers, yet have white petals. Rousseau]; it falls with Chervil into the group with a little involucre and not a large one; its leaves are sufficiently similar to theirs to make it hard to write down for you the differences. But here are some features which will ensure that you do not make a mistake.

One must start by seeing these different plants in flower; for it is in this condition that Hemlock shows its particular characteristics. For beneath each small umbel lies a small involucre of three little pointed leaflets, fairly long, and all three turned outwards, whereas the leaflets of the small umbels on the Chervil completely encircle it, and are evenly turned in all directions. In the case of Parsley, it bears only a few short leaflets, as fine as hair and disposed at random

WILD CARROT *Daucus carota* *a* The flower, seen from the front *b* The fruit

a
b

both on the large umbels and on the small ones, and these leaflets are pale and scant.

When you are quite sure of the Hemlock in flower, you will confirm your identification by lightly crushing and sniffing its foliage; for its stinking and noxious odour will prevent you from confusing it with either Parsley or Chervil, both of which have a pleasant odour. Certain at last that you will not confuse one with the other, you will examine together and separately these three plants in all their stages of development and in all their components, in particular the foliage, which is present more constantly than the flower; and by this examination, compared and repeated until you have achieved certainty at a glance, you will learn to distinguish and recognize the Hemlock infallibly.

Thus does study lead us to the very door of practice, after which practice confers ease on learning.

Draw breath, dear cousin, for this has been an exhausting letter. Moreover I scarce dare promise you more restraint in the one which is to follow; but after that, our path will be lined with flowers. You deserve a garland for the sweetness and perseverence with which you have deigned to follow me through this undergrowth, in no way daunted by the thorns.

---

[1] The sister of Mme Delessert, whom the author called Tante Julie.

[2] Joseph Pitton de Tournefort (1656-1708), French botanist, author of *Éléments de botanique ou Méthode pour connaître les plantes*.

OTHER FLOWER STRUCTURES *a* Umbel: Carrot *b* Solitary flower: Pink *c* Panicle: Millet *d* Cyme: Elder

a

b

c

d

THE SIXTH LETTER

2 May 1773

Although, dear cousin, there is still a great deal wanting to complete our comprehension of our first five families, and I have not always known how to adapt my descriptions to the understanding of our young *Botanophile* (lover of botany), I flatter myself that I have nonetheless given you such an idea of them as to enable you, after a few months of botanical study, to make yourself familiar with the general appearance of each family: so that on seeing a plant, you may guess more or less whether it belongs to one of the five families, and if so, to which; provided always that afterwards, by an analysis of the inflorescence, you verify whether or not you were mistaken in your conjecture. The umbellifers, for example, cast you into a state of some confusion, from which you may recover, whenever it pleases you, through the aid of the details which I have combined with the descriptions: for, after all, Carrots and Parsnips are so common that nothing is more simple at the height of summer than to have one or other of them pointed out flowering in a vegetable garden. Now, a peremptory glance at the umbel and at the plant which bears it will give you such a clear idea of umbellifers that when you first encounter a plant of this family you will rarely be mistaken. This has been my sole aim so far, for we shall not yet become involved in genera and species; and once again, it is not a parrot-like ability to name names that I wish you to acquire, but a true science, and one of the most delightful sciences one could cultivate. I am now moving on to our sixth family, before treading a

DANDELION *Taraxacum officinale* (see page 92)

more methodical path. It may at first confuse you as much if not more than the umbellifers. But my design at present is nothing more than to give you a general notion of the family, especially as we have still plenty of time before the height of its flowering season, and the interval well employed will smooth out difficulties with which we must not yet contend.

Pick one of those little flowers which at this time of the year carpet the meadows, and which around here they call *Daisies*, *Little Marguerites* or simply *Marguerites*. Examine it carefully; for I am sure you will be surprised when I tell you that despite its appearance this flower, so small and delicate, is really composed of two or three hundred flowers, each complete, that is to say that each has its corolla, its seed, its pistil, its stamens, its pollen – in short, each is as perfect in its way as a Hyacinth or a Lily flower. Every one of those little parts, white above and pink beneath, which form a sort of crown round the Daisy, and which seem to you just so many little petals, are in fact true flowers; and each of those little yellow bits which you observe in the centre, and which you perhaps at first assumed to be stamens, are also just as much true flowers. Were your fingers already adept at botanical dissection and were you equipped with a good magnifying glass and an abundance of patience, I would be able to convince you of this truth through the evidence of your eyes; but for the moment you must begin, if you will, by taking my word for it rather than wearying your concentration on minutiae. However, that you may at the least have some idea, remove one of the little white parts from the crown: you will at first think that this little part is flat throughout its length; but study it well where it was attached to the flower, and you will see this end is not flattened but round, and forms a hollow tube, from which sprouts a tiny two-pointed filament; this filament is the forked style of this flower which, as you see, is only flattened at its apex.

DAISY *Bellis perennis*

Now take a look at the yellow bits in the centre of the flower, which I have told you are also flowers in themselves: if the flower is sufficiently advanced, you will see several round the perimeter which are open in the middle, and even divided into several parts. These are monopetalous corollas which are in the process of opening, and with the aid of a magnifying glass you would easily be able to make out within them the pistil and even the anthers by which it is surrounded. Usually the yellow florets in the centre are still rounded and in bud. These like the others are flowers, but they have not yet opened; for they open only progressively, from the perimeter to the centre. That is enough to provide you with a visual demonstration of the possibility that all these components, the white as much as the yellow, are in truth so many perfect flowers; and this is a very consistent phenomenon. Nevertheless, you will see that all these little flowers are packed together and enclosed in one common calyx, that of the Daisy. In regarding the whole Daisy as a single flower, we will then give it a very appropriate name by calling it a *composite flower*. Now, there are a large number of species and genera of flowers like the Daisy, composed of a cluster of other smaller flowers contained within a common calyx. It is these which constitute the sixth family of which I wished to speak to you, that of the *composite flowers*.

Let us begin by avoiding all ambiguity with regard to the word 'flower', reserving that word in the present family for the composite flower, and giving the name *florets* to the little flowers of which it is composed; but let us not forget that, to be accurate, these florets are themselves so many true flowers.

In the Daisy you have seen two types of floret: the yellow ones occupying the centre of the flower and the narrow little white tongues, or straps, which surround them. The former, in their miniature state, are quite similar in shape to the Lily-of-the-Valley

CHINESE ASTER *Aster chinensis*

or Hyacinth flowers, and the latter are somewhat like the flowers of the Honeysuckle. We shall keep the word *florets* for the first group and, to distinguish the second group we shall bestow the name *demi-florets* on them: for they do indeed rather resemble monopetalous flowers which have been trimmed on one side, leaving only a strap which would scarcely form half a corolla.

These two types of floret form different combinations in the composite flower, so that it is possible to divide the whole family into three quite distinct sections.

The first section is formed by those which are only composed of straps [ray florets] or demi-florets, both in the centre of the flower and surrounding the centre: these are called *demi-floret flowers*; and in this section the whole flower is always of a single colour, most commonly yellow. The Dandelion is just such a flower; other examples are the Lettuce, Chicory (which is blue), Scorzonera, Salsify, etc.

The second section comprises the *floret flowers*, those which are made up only of florets and which are also in general of one colour. Examples of these are everlasting flowers, Burdock, Wormwood, Artemisia, Thistle and Globe Artichoke, which is itself a Thistle, whose calyx and receptacle we eat while it is still in bud, before the flower is open or even formed. The 'choke' which we remove from the centre of the Artichoke is nothing but the bunch of florets which are beginning to form, and which are separated from one another by long hairs growing on the receptacle.

The third section is that of the flowers which combine both types of floret. The whole florets invariably occupy the centre, and the demi-florets make up the surroundings or circumference, as you have seen in the Daisy. The flowers in this section are called *radiate*, since botanists have given the name *rays* to the outer parts of a composite flower when they are formed of straps or demi-florets. As

CHICORY *Cichorium intybus*

regards the space in the centre of the flower, which is filled with the florets, this is called the *disk*; and they also sometimes give this same name 'disk' to the surface of the receptacle from which grow all the florets and demi-florets.

In radiate flowers, the disk is often of one colour and the rays of another; however, there are also genera and species where both are of the same colour.

Let us now endeavour to clarify your notions of a *composite flower*. The ordinary Clover is now in flower; its flower is purple: if you should come across one, you might, on seeing so many little flowers gathered together, be tempted to take the whole for a composite flower. You would be wrong. Why? Because a composite flower is not merely an aggregation of several small flowers; it must, in addition, have one or two parts of the inflorescence in common to all the florets, so that they all share these same parts, and none separately possesses its own. These two parts are the calyx and the receptacle. It is true that the flower of the Clover, or rather the grouping of flowers which seem but one, at first appears to be borne on a sort of calyx; but pull this so-called calyx a little to one side, and you will see that it is not touching the flower at all, but is attached below it to the pedicel on which the flower is carried. Thus this apparent calyx is not a calyx at all; it is a part of the foliage and not of the flower; and this so-called flower is really no more than a collection of very small leguminous flowers, each with its very own calyx, with absolutely nothing in common save that they are attached to the same pedicel. It is customary, however, to consider it all as one single flower; but that is a false assumption; or if one insists on regarding such a cluster as one flower, one must at least call it not a *composite flower*, but an *aggregate flower*, or a flowerhead (*flos aggregatus*, *flos capitatus*, *capitulum*). And these terms are indeed sometimes used in this sense by botanists.

COMPOSITE FLOWERS *a* Demi-floreted *b*, *i* Floreted *c* Rayed *d* Single male floret *e* Single female demi-floret *f* Sterile floret *g* Hermaphrodite demi-floret *h* Hermaphrodite floret *k* Floret with ovary and pappus *l* Floret *m* Stalked pappus *n* Sessile pappus *o* Scale from common calyx *p* Receptacle

a
b
i
c
d
e
f
g
h
k
l
m
n
o
p

There you have, dear cousin, the simplest and most straightforward idea of the family that I can give you, or rather of the large class of composites, and of the three sections or families into which it is subdivided. I must now tell you about the structure of the inflorescences peculiar to this class, and that will perhaps help us more accurately to determine its particular nature.

The most important part of a composite flower is the receptacle, on which grow first the florets and demi-florets and later the seeds. This receptacle, a sizable disk, lies in the centre of the calyx, as you see in the Dandelion, which we shall take as our example. The calyx, throughout the family, is normally divided almost to the base into several pieces, enabling it to close up, open out and fold back, as occurs during the stages of fruiting, without thereby causing it to tear. The calyx of the Dandelion is made up of two tiers of leaflets, one within the other; and these leaflets in the outer tier that supports the inner, curve outwards and downwards towards the pedicel, while the leaflets of the inner tier remain upright, encircling and holding in the demi-florets which form the flower.

One of the most common forms of calyx in this class is called *imbricate*, that is, formed of several tiers of leaflets, each overlapping the next, like tiles on a roof. The Artichoke, Cornflower, Knapweed and Scorzonera are examples of species with imbricate calyxes.

The florets and demi-florets contained by the calyx are massed very densely on the disk or receptacle, staggered like the squares of a draught-board. Sometimes they are in unimpeded contact with each other, nothing intervening; at other times they are separated by barriers of hairs or by tiny scales which remain attached to the receptacle when the seeds have fallen. Now you are already on the way to noticing the differences in the calyxes and receptacles; let us now discuss the structure of the florets and demi-florets, beginning with the former.

CALYX AND REPRODUCTIVE PARTS (see Dictionary) *a* Perianth: Pink *b* Glumes and awns: Grasses *c* Spathe and spadix: Arum *d* Stamen: Lily *e* Pistil: Lily *f* Floral disk: Buckthorn

a
b
c
d
e
f

A floret is a monopetalous flower, usually regular, with a corolla split at its top into four or five parts. The filaments of the stamens, five in number, are attached to the inside of the corolla tube: these five filaments are conjoined at their tips into a small, circular tube which encircles the pistil, and this tube is nothing more than the five anthers or stamens united in a circle to form a single entity. This uniting of the stamens constitutes, in the eyes of botanists, the essential characteristic of composite flowers, and is found only in their florets and in no other type of flower. Thus, even should you find several flowers borne on the same disk, as in the Scabious and the Teasel, if the anthers are not united in a tube round the pistil, and if the corolla is not growing from a single naked seed, then these flowers are not florets and do not constitute a composite flower. On the other hand, if you should find in a single flower the anthers thus united to form an entity, and the superior corolla growing from a single seed, then this flower, even though single, would be a true floret, and would belong to the composite family, whose character is better ascertained from a precise structure than from a misleading appearance.

The pistil possesses a style which is usually longer than the floret, above which one sees it growing through the tube formed by the anthers. It is most often terminated at its tip by a forked stigma whose two little horns one can easily see.

At its base, the pistil does not grow directly from the receptacle and nor does the floret, for both are secured to it by the seed, which acts as their anchor, and which grows and lengthens as the floret withers, and finally becomes an elongated seed which remains attached to the receptacle until it is ripe. Then, if it is naked, it falls, or if it is crowned with a tuft of fluff the wind bears it away, and the receptacle is left, nakedly exposed in some genera, covered with scales or hairs in others.

SWEET SCABIOUS *Scabiosa atropurpurea*

The structure of the demi-florets is similar to that of the florets; the stamens, the pistil and the seed are organized in much the same way: except that in the radiate flowers there are several genera in which the demi-florets round the outside tend to be fruitless, either because they lack stamens, or because those they have are sterile, and have not the power to fertilize the seed; then the flower sets seed only among the florets in the centre.

Throughout the composite class, the seed is always *sessile*, which means that it lies directly on the receptacle without any intervening pedicel. But there are some seeds whose tip is surmounted by a tuft, which may be sessile or may be attached to the seed by a pedicel. You understand that the purpose of this tuft is to scatter the seeds abroad by catching the wind, which carries them off and sows them far and wide.

To these rough and abbreviated descriptions I should add that the calyxes are usually designed to open out when the flower blooms, and to close up again when the florets seed and fall, thus retaining the young seed, and preventing it from being scattered before it has matured; finally the calyxes open out again and fold right back, creating a larger space in the centre for the seeds, which swell as they ripen. You must often have seen the Dandelion in this state, when children pluck it to blow off the tufts which form a globe round the reflexed calyx.

To know this class well, one must follow the flowers from before their opening until the time when the fruit is fully mature; and it is in this series of events that one sees the transformations and succession of marvels which hold any alert observer in continual admiration. A flower which lends itself to such observation is the Sunflower, which one can often discover in vineyards and in gardens. The Sunflower, as you see, is a radiate composite. The Marguerite, which graces the autumn flower borders, is also a radiate. Thistles

CORNFLOWER *Centaurea cyanus* (see page 96)

are floreted flowers; I have already said that Scorzonera and Dandelion are demi-floreted. All these flowers are large enough to be dissected and studied with the naked eye without too much effort.

I shall not tell you any more today about the family or class of composites. I greatly fear that I have already abused your patience with details which would have been clearer had I known how to make them briefer; but I cannot avoid the difficulties which arise through the small size of the objects. Farewell, dear cousin.

COROLLAS *a* Inferior flower: Hound's-tongue *b* Superior flower: Campanula *c* 'Rosette' flower: Elder *d* Petal of Pink *e* Flower of St John's Wort *f* Spurred 'mouth' flower: Toadflax *g* Petal of Cistus

a
b
c
d
e
f
g

THE SEVENTH LETTER

Of Fruit Trees

(undated; probably spring 1774)

I was awaiting news of you, dear cousin, but without impatience, since Monsieur T., whom I have seen since I received your last letter, told me that he had left your mama and all your family in good health. I am delighted to have confirmation of this from yourself, and also to have the recent good news from you of my aunt Gonceru.[1] Her blessing and good wishes have filled with joy a heart that for long has scarcely experienced such emotions. In her I feel that I still have something very precious in this world; and as long as I still have her I shall continue, whatever may happen, to love life. Now is the time to take advantage of your usual kindness to her and to me; it seems to me that my little offering[2] acquires real value when it passes through your hands. If your beloved husband is coming to Paris soon, as you lead me to hope, I shall ask him to be obliging enough to carry my annual tribute; but should he be delayed a little, please let me know to whom I should entrust it, so that it is not late, and so that you do not make me a loan of it as you did last year: I know you do it gladly, but I cannot allow it when it is not necessary.

Here, dear cousin, are the names of the plants you recently sent me. I have added a question mark to those of which I am uncertain, because you neglected to accompany some of the flowers with their leaves, and in determining species, leaves are often essential for

PEACH BLOSSOM *Prunus persica* (see page 110)

such a poor botanist as myself. When you arrive at Fourrière, you will find most of the fruit trees in blossom, and I recall that you requested some guidance on this subject. As I am in a great hurry, I can only scribble you a few very hasty words for the present, in order that you should not lose yet another season for this study.

You must not, my dear friend, give botany an importance which it does not have; it is a study of pure curiosity, one that has no real utility except what a thinking, sensitive human being can draw from observing nature and the marvels of the universe. Man has rendered many things unnatural, the better to put them to his own use: in that he is in no way to be blamed; but it is nevertheless true that he has often disfigured them, and that when he believes he is truly studying nature in the works of his own hands, he deceives himself. This error is particularly prevalent in civilized society; it even occurs in gardens. Those double flowers admired in the flowerbeds are monsters deprived of the faculty of reproducing their kind, a faculty with which nature has endowed all living things. Fruit trees are in much the same situation by being grafted: you will plant the pips of the finest varieties of Pears and Apples in vain: nothing will grow from them but wildings. Thus, to appreciate the Pear and the Apple as nature fashioned them, you must seek them not in kitchen gardens but in the forests. Their flesh is not so plump and juicy, but the fruits ripen better, and they multiply in greater numbers, and the trees are far taller and more vigorous. But I am broaching a subject here which would lead me too far: let us return to our kitchen gardens.

Our fruit trees, although grafted, retain in their inflorescences all the botanical characteristics which distinguish them; and it is through careful study of these characteristics, as well as of the transformations wrought by grafting, that one realizes, for example, that there is only one species of Pear, under a thousand different names

PEACH FRUIT *Prunus persica* (see page 110)

which have divided it according to the shape and flavour of the fruit into so many so-called species, which in reality are nothing but varieties. Further, the Pear and the Apple are but two species of the same genus; and the only really characteristic difference between them is that the pedicel of the Apple is embedded in the fruit, while that of the Pear grows from a somewhat elongated protuberance of the fruit. In the same way, all the varieties of Cherries: Heart Cherries, Morello Cherries, White Heart Cherries are merely varieties of one species; all Plums constitute only one species of Plum. The genus *Prunus* comprises three main species: namely, the true Plum, the Cherry, and the Apricot, which is also only a species of Plum. Thus, when the learned Linnaeus divided the genus into its species, and called them *Prunus*-Plum, *Prunus*-Cherry, and *Prunus*-Apricot, ignorant people ridiculed him; but those who looked closely admired the correctness of his classification, etc. We must move on; I make haste.

Almost all fruit trees belong to one large family, whose characteristics are easy to grasp: the stamens, which are numerous, instead of being attached to the receptacle are attached to the calyx in the gaps left between the petals; all the flowers are polypetalous and usually have five petals. These are the main generic characteristics: The Pear genus, which also includes the Apple and the Quince. Five-pointed calyx in one piece. Five-petalled corolla attached to the calyx; about twenty stamens all attached to the calyx. The seed or ovary inferior, that is to say beneath the corolla; five styles. Fruits fleshy with five compartments containing the seeds, etc.

The Plum genus, which includes the Apricot, the Cherry and the Cherry-Laurel. Calyx, corolla and anthers all more or less like the Pear. But the seed is superior, that is to say inside the corolla, and there is only one style. Fruit watery rather than fleshy, containing one stone, etc.

PLUM FRUIT *Prunus domestica*

The Almond genus, which includes the Peach. Almost as the Plum, except that the seed is velvety, and the fruit, soft in the Peach, dry in the Almond, contains a hard kernel, rough and covered with fissures.

All this is only very roughly outlined, but it is enough to amuse you this year. Farewell, dear cousin.

---

[1] Suzanne Goncerut, who looked after the young Jean-Jacques for some years after the death of his mother.

[2] Rousseau sent his aunt an annual gift of money in gratitude for what she had done for him.

PLUM BLOSSOM *Prunus domestica*

THE EIGHTH LETTER

On Pressed Plant Collections

11 April 1773

Heaven be praised, dear cousin, for your recovery. I was already quite alarmed by your silence and by that of Monsieur G., whom I had earnestly begged to send me word of his arrival. When one is worried in this way, nothing is more cruel than silence, because it makes one fear the worst. But all that is now forgotten, and I feel nothing but pleasure at your recovery. The return of the fine weather, your less sedentary existence at Fourrière, and the joy of successfully fulfilling the sweetest and most worthy of duties will soon complete it; and you will feel less sadly the temporary absence of your husband, surrounded as you are by cherished tokens of his affection who demand so much of your attention.

The earth is beginning to turn green, the trees to bud, and the flowers to open; some are already over: a moment's delay in our botanical studies would defer them for a whole year; so I proceed without further preamble.

I fear that we may have treated our studies up to now in too abstract a fashion, by not applying them to specific objects. This is the error into which I have fallen particularly in the case of the umbellifers. If I had begun by showing you one, I would have spared you a most wearying study of an imaginary object, and myself some difficult descriptions which one look at the plant would have rendered unnecessary. Unfortunately, at the distance imposed on me

HONESTY *Lunaria rediviva* (see page 42)

by force of circumstance, I am not in a position to point out the plants to you; but if each of us independently can have the same examples under our eyes, we shall comprehend each other very well when we discuss what we are looking at. The only problem is that the description must come from you; for to send you dried plants from here would be pointless. To recognize a plant properly, one must first see it growing. Dried plant collections are useful as reminders of those one already knows, but they are very little use for learning to know those one has not seen before. It is, therefore, for you to send me plants which you would like to know about and which you have gathered as they grow; my part is to name them, classify them and describe them for you, until your eyes and your mind have become so accustomed to the techniques of comparison that you yourself are able to classify, order and name plants which you see for the first time: this is the knowledge which alone distinguishes the true botanist from the herbalist or taxonomist. It is a question, then, of learning how to prepare, dry and preserve plants or parts of plants, in such a way as to make them easy to recognize and name. In a word, I propose that you start a pressed flower collection. This is the beginning of a great undertaking for our little amateur botanist to take over in the future: for now, and for some time to come, the nimbleness of your fingers must compensate for the clumsiness of hers.

First you must make some purchases: five or six quires of grey paper, and about the same quantity of white of the same size, quite strong, and well made, else the plants would rot into the grey paper, or at least the flowers would lose their colour, which is one of the features by which you can recognize them and which makes a pressed flower collection attractive to the eye. It would also be desirable for you to have a press of the same size as your paper, or, failing that, two very smooth pieces of wood, so that by inserting

COMMON SCURVY-GRASS *Cochlearia officinalis* (see page 42)

your leaves between them they can be kept pressed by stones or other heavy objects with which you will weigh down the top piece of wood. When you have completed these preparations, this is what you must do to prepare your plants so that you will be able to preserve and recognize them.

The moment to select for this purpose is that at which the plant is in full bloom, and when a few flowers are beginning to fall to make room for the fruit which is beginning to appear. It is at this point, when all the stages of the fruiting process are apparent, that you must try to pluck the plant and dry it in this condition.

Small plants can be gathered complete with their roots, which you clean carefully with a brush, to ensure that no earth remains on them. If the earth is damp, let it dry before brushing it off, or else wash the root; but in this case you must take the greatest care to wipe it well and dry it before placing it betwen the papers, otherwise it would undoubtedly rot there, and pass on its rot to the surrounding plants. It is, however, only important to preserve the roots when they possess some noticeable peculiarities; for in most cases, the branching and fibrous roots are so similar in form that it is not worth the trouble of preserving them. Nature, which has lavished so much elegance and decoration upon the shape and appearance of the visible parts of the plants, has ordained for the roots only useful functions since, hidden as they are under the earth, to give them an attractive appearance would have been to hide their light under a bushel.

In the case of trees and all the large plants, only fragments are taken. But this fragment must be so well chosen that it carries all the constituent parts of the genus and species, so that it is complete enough to enable one to recognize and identify the plant which has provided it. It is not enough that all parts of the inflorescence are perceptible; that will only serve to identify the genus. One must

WHITE HOREHOUND *Marrubium vulgare* (see page 64)

clearly see in the fragment the character of the foliation and ramification, which is to say the disposition and shape of the leaves and twigs, and even, where possible, some parts of the stem: for, as you will see later, all these are important in the identification of the different species of the same genus, which are identical in flower and fruit. If the twigs are too thick one reduces them with a knife or penknife by carefully paring them down on one side as far as one can without cutting and damaging the leaves. There are botanists who have the patience to split the bark of the twig and neatly withdraw the wood, so that when the bark is reassembled it still seems to represent the whole twig, even though the wood is no longer there, by which means one avoids excessive bumps and unevennesses between the papers, which spoil and disfigure the collection, and deform the plants. In the plants where flowers and leaves do not appear at the same time, or where they grow too far apart from each other, one takes a small twig with flowers and a small twig with leaves and, placing these together on the same piece of paper, one thus presents the various constituents of the same plant, in a way which makes it recognizable. As for plants where one finds only leaves and where the flower has not appeared or is already over, one must leave them, and wait until they show their face, in order to recognize them. A plant is no more certainly distinguishable by its leaves than a man by his clothes.

Such is the choice that must be made of what one picks: one must also choose the moment for doing this. Plants picked in the morning dew or in the dampness of evening or when it is raining during the day do not last. It is quite essential to choose dry weather, and even in such weather it must be the driest and hottest time of day, which in summer is between eleven o'clock in the morning and five or six o'clock in the evening. Even then, if the slightest moisture is on them, they must be left; for they will never keep.

SHEPHERD'S PURSE *Capsella bursa-pastoris* (see page 42)

When you have picked your specimens, take them home, keeping them quite dry, and lay them out and arrange them on your papers. To do this you build up a layer of at least two sheets of grey paper, on which you place one sheet of white, and on this sheet you arrange your plant, taking great care that all its parts, especially the leaves and flowers, are wide open and spread out in their natural position. A plant which is slightly withered, though not too much, is usually more amenable to arrangement, which you execute with finger and thumb on the paper. But some rebel, twisting to one side as you are arranging the other. To prevent this annoyance, I keep weights, heavy coins by which I hold down the parts I have just arranged while I arrange the others; so that when I have finished, my plant is almost completely covered with these weights which hold it in position. After this, one places a second white sheet of paper on the first and presses it with the hand, to keep the plant firmly in position: pressing down in this way with the left hand, one moves it forward, while at the same time removing with the right hand the weights and coins which lie between the papers; then one places two more sheets of grey paper on the second sheet of white paper, without for a moment releasing one's hold on the positioned plant, to make sure that it does not slip out of position: on this grey paper one lays a further white sheet, on this sheet another plant, which is arranged and covered as before, until everything one has gathered has been laid out; there should not be too many on each occasion, so that your task may not be too drawn out and that your paper may not become too damp during the drying of the plants because there are so many of them; this would certainly spoil the plants were you not quickly to change the paper, taking the same precautions as before; and indeed you must do this from time to time, until the plants have settled into position and are all sufficiently dry.

TUBEROSE *Polianthes tuberosa* (see page 32)

Your pile of plants and sheets of paper, so arranged, should be put into a press, failing which the plants would curl up: some need to be pressed for a long time, others less; experience will teach you this and also to change the paper wherever and whenever necessary without giving yourself inessential work. Then, when your plants are thoroughly dry, you will place them neatly on their own piece of paper, one above the other; there is no need for further paper in between; and you will thus have the beginnings of a collection, which will continually increase with your knowledge and will eventually contain the story of all the vegetation of the area. Finally, one must always keep a plant collection tightly closed and lightly pressed; otherwise the plants, however dry they are, would attract moisture from the air and would again curl up.

Here now is the way in which to go about this task so that you arrive at a detailed knowledge of the plants, and so that we clearly understand each other when we discuss them.

You must pick two specimens of each plant; the larger one is for you to keep and the smaller one is to be sent to me. Number them carefully, so that the large and small specimens of each species always have the same number. When you have one or two dozen species dried in this way, send them to me in a small notebook when the opportunity arises. I shall send you the name and description of those same plants: with the help of the numbers you will identify them in your collection and then in the field, where I am assuming you will have started to study them closely. This is a certain way to make your progress as sure and rapid as possible so far away from your mentor.

N.B. I forgot to tell you that the same papers may be used several times, as long as you are careful to air and dry them beforehand. I must add that the collection must be stored in the driest place in the house, preferably on the second rather than the ground floor.

YELLOW ASPHODEL *Asphodelus lutea* (see page32)

NOTES TOWARDS A DICTIONARY OF BOTANICAL TERMS

Editorial note

Rousseau left only rough drafts of what appears to have been an attempt to produce a Dictionary which would provide the layman with an explanation of the terms familiar to botanists. It would be impossible, without copious footnotes, to present them in as complete a form as he left them, and at the same time avoid the danger of misleading the reader.

Thus it has seemed better in this edition to omit those entries in which Rousseau was completely mistaken or where his terminology was never adopted or has fallen into disuse. Also, on several occasions editorial notes have been inserted in square brackets either to rectify Rousseau's statements or to add to them later discoveries of which he was unaware.

In certain fields botanical knowledge was well advanced in the eighteenth century; in others, through lack of study or essential equipment, it was in its infancy; and Rousseau himself was often curiously unwilling in his Dictionary to employ terms of which from the evidence of his other writings he was perfectly well aware but for some reason did not avail himself. (An example is *poussière prolifique*, when the word *pollen* was widely used.) However, he also evinces in many of the entries a remarkable prescience far ahead of current botanical thought – his dismissal of the Linnean analogy of the parts of Mosses is a case in point.

In this edition the Dictionary has been arranged alphabetically under the English equivalents, with the French terms used by Rousseau set afterwards in brackets. Some of the nuances are lost in translation, but in general this is because Rousseau was using unnecessarily or incorrectly different terms for structures or concepts which are no longer felt to be valid.

In the long and sometimes rhapsodical entry on 'Flower', the section describing composite flowers has been omitted, since it is largely repeated in the Sixth Letter.

ACAULOUS (Acaulis) Stemless.

ACCESSORY PARTS (Supports, *Fulcra*) There are ten types; namely, the stipule, the bract, the tendril, the spine, the thorn, the pedicel, the petiole, the stem, the gland and the scale.

ANDROGYNOUS (Androgyne) Bearing male and female flowers on the same plant. The words *androgynous* and *monoecious* mean absolutely the same thing, except that in the former case one is referring rather to the separate sexes of the flowers, and in the latter to their presence on the same plant. [Rousseau was incorrect. Androgynous refers only to male and female elements in the same flower; monoecious to separate male and female flowers on the same plant.]

ANGIOSPERM (Angiosperme) Plant with seeds completely enclosed in an ovary. This term refers both to fruits in capsules and to berries.

ANTHER (Anthère) The capsule or compartment carried at the end of the filament of the stamen. At the moment of fertilization it opens and scatters the numerous pollen grains.

GERMINATION *a* Leaf bud: Chestnut *b* Flower bud: Peach *c-f* Beans at various stages of germination *g*, *h* Young Beech plants *i*, *k*, *l* Wheat at various stages of germination

a
b
c
d
e
f
g
h
i
k
l

ARTICULATED, JOINTED (Articulé) A term applied to stems, roots, leaves and seed vessels when any of these parts of the plant is divided at intervals by nodes (see NODES).

AXIL (Aisselle) Acute or right angle between one branch and another, or a branch and the trunk, or a leaf and a branch.

AXILLARY (Axillaire) Arising from an axil.

BARK (Écorce) The exterior of the trunk and of the branches of a tree. The bark lies between the epidermis on the outside and the bast or interior bark. These three layers are usually grouped under the common name 'bark'.

BAST, PHLOEM (Le Liber) This is composed of thin layers, rather like the leaves of a book; they are in direct contact with the wood. The bast breaks away each year from the other parts of the bark, and merges with the sapwood to create a new ring round the circumference of the tree, so increasing the diameter of the trunk.

BERRY (Baie) Fleshy or juicy fruit, with one or more seed chambers.

BRANCHES (Branches) Supple and springy arms of the main part of the tree. It is these which give the tree its shape. They may be disposed alternately, opposite or spirally. The leaf bud slowly grows into subsidiary branches, which arise from the same part of the main stem; and it is claimed that the stirring of the branches by the wind is to trees what the beat of the heart is to animals. Branches may be divided into the following categories:

1 Primary branches, springing directly from the trunk, and from which all the other branches stem.
2 Woody branches, which are the thickest and are packed with flat buds. These give the shape to a fruit tree, and to some extent protect it.
3 Fruiting branches, which are the weakest, and bear round buds.
4 Twiggy branches, which are thin and short.
5 Lazy branches, which are thick, long and straight.
6 Weak branches, which are long and unproductive.
7 August branches, which after the advent of August harden and become blackish.
8 Finally, the false-wood branch, which is thick where it should be thin, and shows no signs of fertility.

[Few of these terms are used today.]

BULB (Bulbe) A spherical root, composed of several layers or tunics fitting closely round each other. Bulbs are underground buds rather than roots. They themselves have roots, which are normally almost cylindrical and branched.

BULBILS (Cayeux) Adventitious bulbs by which several Lilies and other plants reproduce themselves.

CALYX (Calice) External covering or protection for other parts of the flower, etc. Just as there are some plants with no calyx at all, so there are some whose calyx is gradually metamorphosed into the leaves of the plant, and conversely there are plants whose leaves change into a calyx. This can be seen among the Ranunculaceae in plants such as the Anemone and the Pasque Flower.

CAPILLARIES (Capillaires) Capillary leaves are those found in the Moss family which are as fine as hairs.

The name 'Capillaires' is also given to a branch of the Fern family [Maidenhair-fern *Adiantum capillus-veneris*], whose sori are found on the back of its leaves, as in other Ferns, and which is distinguished from them only by its much smaller size.

CAPRIFICATION (Caprification) The fertilization of the female flowers of a certain type of dioecious Fig by pollen from the stamens of the male specimen, named the Wild Fig. Through this natural process, helped by human effort, the figs thus fertilized swell and ripen, and produce a far greater crop than would otherwise have been obtained.

What is remarkable in this sequence is that in the Fig genus the flowers are enclosed in the

DETAILS OF PEAR *Pyrus communis* (see page 106)

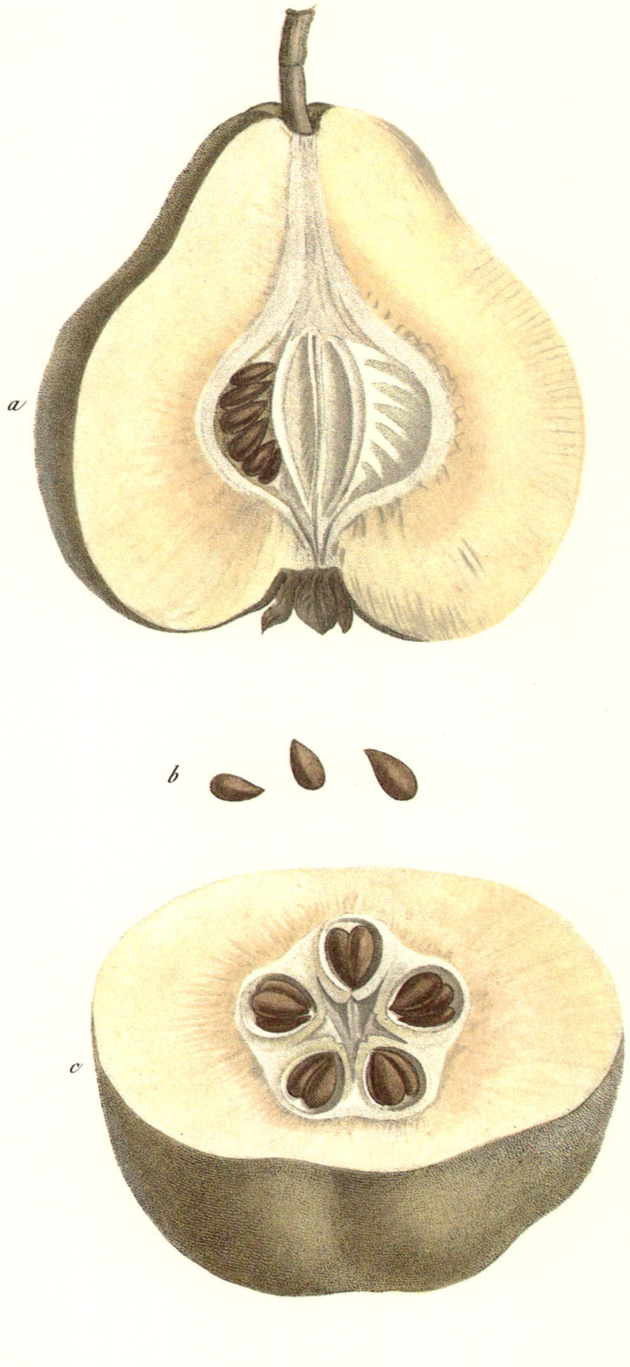
a
b
c

fruit, and it is only those flowers which are hermaphrodite or androgynous which seem capable of being fertilized. For when the sexes are entirely separated, one cannot imagine how the pollen from the male flowers can penetrate both its own covering and that of the female flowers to reach the pistils which it should fertilize.

It is an insect which undertakes this mission. A type of gall-insect peculiar to the Wild Fig lays and hatches its eggs within the receptacle, and is thus covered by the pollen from the stamens, which it carries through the bud of the fig, past the scales which guard the entrance, right to the interior of the fruit. There, this pollen, encountering no obstacle, is deposited on those organs which are fashioned to receive it.

CAPSULAR (Capsulaire) Capsular plants are those in which the fruit is enclosed in capsules or pods.

CAPSULE (i) (Capsule) Dry pericarp of a dry fruit. One would not, for instance, give the name capsule to the rind of the pomegranate, for though it is as dry and hard as many other capsules, it encloses a soft fruit.

(ii) (Urne) Case or capsule filled with spores, borne by most mosses in flower. The commonest structure of these capsules is carried above the plant by a pedicel of variable length; at their top they have a sort of pointed cap or hood which covers them; attached at first to the capsule, it later detaches itself and falls when the capsule is ready to open. The capsule then opens two-thirds of the way up, like a soap box, by means of a lid which detaches itself and falls in its turn after the cap has fallen. It has a double fringe of hairs where it joins, so that moisture cannot penetrate the capsule as it would when open. Finally, as maturity approaches, it bends over and curves downwards to tip the spores which it contains on the ground.

The general opinion of botanists on this subject is that the capsule and its pedicel are a stamen; the pedicel is the filament, the capsule is the anther, and the spores, which it contains and releases, are the pollen which fertilizes the female flower; as a result of this explanation, the name anther is often given to the capsule which we are discussing. However, as the fruiting system of mosses is not at the moment perfectly known, and as there is no unshakable certainty that the anther we are discussing is really an anther, I believe that pending more impressive evidence, and without hastening to adopt so definite a name (which greater knowledge might later force us to abandon), it is better to use the word capsule, which Vaillant [Sébastien Vaillant (1669-1722), botanist, author of *Botanicon Parisiense* (1727), who became director of the Jardin des Plantes] gave it, and which can be conveniently kept, whatever system is adopted.

CARPEL (Loge) Cavity inside a fruit: when divided by partitions, there are several cells.

CATKIN (Chaton) Structure of male or female flowers spirally attached to an axis or common receptacle, around which the flowers take on the form of a cat's tail. There are more trees bearing male catkins than female ones.

CLAW (Onglet) A sort of hooked point by which the petal in some corollas is attached to the calyx or to the receptacle: the claw of Pinks is longer than that of Roses.

COMMON (Vulgaire) Name usually given to the main and oldest known species, from which the genus takes its name, and which was first regarded as the only species. [This is not correct. 'Common' (*vulgaris*) is applied to the most commonly found species.]

CORYMB (Corymbe) Flower form between an umbel and a panicle. As in a panicle, the pedicels grow shorter towards the top, and all reach the same level, creating a level surface at the top. The corymb differs from the umbel in that the pedicels, rather than radiating from the same point, diverge at different heights from various points on the same axis.

COTYLEDON (Cotyledon) A small leaf, part of the seed, in which the nutritious saps of the new plant are prepared and built up.

The cotyledons, otherwise known as seed-leaves, are the first part of the plant to be seen above the ground after germination has started.

PEAR BLOSSOM *Pyrus communis* (see page 106)

a
b

These primary leaves are often of a different shape from those that follow, which are the true leaves. Usually the cotyledons soon wither, and they fall shortly after the plant has appeared. The plant then receives nourishment from other quarters in greater quantity than that which it would draw through the cotyledons from the substance of the seed itself.

Some plants have only one cotyledon, and so are called *monocotyledons*: among them are Palms, Lilies, Grasses and other plants. The largest number have two cotyledons, and are called *dicotyledons*. If others have more, they will be called *polycotyledons*. *Acotyledons* are those with no cotyledons at all, such as Ferns, Mosses, Fungi and all cryptogams (plants with no stamens or pistils).

But in order to classify plants in this way, they must be observed as they emerge from the ground, and in the very seed itself. This is often very difficult to do, especially in the case of marine and aquatic plants, and with trees and foreign or alpine plants which refuse to germinate and grow in our gardens.

CREEP (Tracer) To have creeping rhizomes, like Couchgrass. Thus the word only applies to roots. When one says that the Strawberry creeps, one is wrong: it ramps, which is different.

CRUCIFEROUS or CRUCIFORM (Crucifère ou Cruciforme) Arranged in the form of a cross. The name Crucifer is given in particular to one family in which the flowers are composed of four petals arranged in the form of a cross, on a calyx similarly composed of small leaves, with six stamens grouped round the pistil, two of which, while of the same length as the other, are shorter than the other four and separate them into two sets.

CULM (Chaume) A name which specifically distinguishes the jointed and hollow stem of Grasses from that of other plants. While many other plants possess this same feature, it is not shared by Sedges.

One may add that the culm is never branched; though there are exceptions such as in the *Arundo calamagrostis* [*Calamagrostis lanceolata*] and in some other species.

CUPS (Cupules) Types of small skull-caps or cups which grow most often on a variety of Lichens and Algae, and in the bowl of which one can see the seeds growing and forming, particularly in the Liverwort genus *Marchantia*. [Not used in this sense today.]

CUTTING (Bouture) A young stem which is cut from certain softwood trees, such as Fig, Willow and Pear, and is planted in the soil with no roots. The success of cuttings depends more on their ability to produce roots than on the softness of their stems. For the Orange, Box, Yew and Juniper, which have little softwood, grow easily from cuttings.

CYME (Cyme ou Cymier) Type of inflorescence which has no regular features, even though all the spokes radiate from the same point. Examples are the Guelder Rose and Honeysuckle. [Incorrect. A cyme is an inflorescence whose growing points are each terminated by a flower.]

DEFECTIVE FLOWER (Fleur mutilée) This is a flower which, usually because of a lack of heat, loses or does not produce at all the corolla which it should naturally possess. Although this abnormality does not constitute a species, plants in which this takes place are nevertheless separated in nomenclature from those of the same species which are complete, as one can see in several species of Morning Glory, Catchfly, Coltsfoot and Bellflower, etc.

DEMI-FLORET (Demi-Fleuron) Tournefort [Joseph Pitton de Tournefort (1656-1708), botanist, author of *Éléments de botanique ou Méthodes pour connaître les plantes* (1694)] gives this name to the toothed florets of Compositae which make up the disk in Lettuces, and to those which make up the outer florets of composite flowerheads. Although these two types of demi-florets are of exactly the same shape, and for that reason are grouped together by botanists under the same name, they nevertheless have essential differences, in that the former always have stamens, whilst the latter

FENNEL *Foeniculum vulgare* (see page 82)

never do. Demi-florets, like florets, are always superior, springing from seeds which in turn grow from the disk or receptacle of the flower. The demi-floret is made up of two parts: the lower is a very short tube or cylinder, and the upper is flat, strap-like, and gives the demi-floret its name. (see FLORET, FLOWER.)

DIGITATE (Digité) A leaf is digitate (palmate) when all its leaflets arise from the tip of the petiole as from one central point. A typical example, for instance, is the Horse Chestnut.

DIOECIOUS (Diécie ou Dioecie) Plants living in separate houses. The word dioecious is applied to plants among which all the male flowers are carried on one stock and the female flowers on another.

DISK (Disque) The intermediate part which supports the flower or some of the upper parts above the true receptacle.

Sometimes, as in the Compositae, the disk is the receptacle itself. In this case one makes a distinction between the surface of the receptacle or disk and the border which surrounds it, which is called the radius.

A disk is also a fleshy structure which is found in some plant genera at the base of the calyx, below the seed. Sometimes the stamens are attached to the edges of this disk.

EAR or SPIKE (Épi) Formation of an inflorescence in which the flowers are attached round a common axis or receptacle formed by the tip of a single stem or stalk. When the flowers are pediculate [stalked], as long as the stalks are single and attached closely to the axis, the inflorescence is always called a spike [correctly a raceme]; but in a true spike the flowers are sessile [stalkless].

EPIDERMIS (l'Épiderme) This is the thin exterior skin which covers the cortical layers. It is a very fine, transparent membrane, normally colourless, elastic and slightly porous.

EYE (Ombilic) In berries and other *inferior* soft fruit, the eye is the receptacle of the flower; its scar remains on the fruit, as one can see on Bilberries. Often the calyx persists and surmounts the eye, which is then commonly called the 'top'. And so it is on Pears and Apples, where the dried up calyx persists round the eye.

EYES (Oeilletons) Buds on the sides of the roots of Artichokes and other plants, which one detaches to propagate these plants.

FERTILIZATION (Fécondation) [Pollination] Natural process through which the stamens carry to the ovary, by means of the pistil, the life force essential for the ripening of seeds and their germination.

FILAMENT (Filet) Short stalk which supports the stamen. The word 'filet' is also given to fine hair on the surface of leaves, stalks and even on the flowers of several plants. [The latter definition is not used.]

FLORET (Fleuron) Small, incomplete flower found in the composition of a composite flower head. See FLOWER [see the Sixth Letter].

This is the normal structure of a composite floret:

1 Monopetalous tubular corolla with five teeth, superior.
2 Elongated pistil, terminated by two reflexed stigmas.
3 Five stamens whose filaments are separated at the base, but which form a tube round the pistil by the conjunction of the anthers.
4 Elongated, naked seed, with the receptacle as its base, itself providing at its tip the receptacle for the corolla.
5 Tuft [pappus] of fine hairs or scales topping the seed and representing a calyx at the base of the corolla. This tuft pushes the corolla upwards from beneath, detaches it and causes it to fall when it has withered, and when the swollen seed is approaching maturity.

This generally common structure of the florets allows exceptions in several genera of the Compositae, and these differences really make up the sections which form the many branches of this large family.

INDIGO *Indigofera tinctoria* (see page 54)

Those differences which pertain to the very structure of the florets have been explained above under FLOWER. [See the Sixth Letter.] I must now discuss those which are concerned with reproduction.

The normal state of the florets of which I have just spoken is hermaphrodite, and they are self-fertilized. But there are others which bearing stamens but no seed are called male; others which have a seed and no stamens are called female florets; others with neither seed nor stamens, or whose imperfect seed always fails, are called neuter.

These various types of floret are not scattered at random among the composite flowers; but their orderly, regular arrangement is always related either to the best means of ensuring fertilization, or to the most prolific fruiting, or to fullest ripening of the seeds.

FLOWER (Fleur) If I should let my imagination surrender to the sweet sensations which this word seems to evoke, I would be able to write an article pleasing perhaps to shepherds, but most unacceptable to botanists. So for a moment let us forget bright colours, sweet scents, elegant shapes, to discover first how to know really well the organism which embraces all these properties. Nothing at first would seem more easy. Who feels the need to know what a flower is? When no one asks me the time I know it perfectly well, as St Augustine said; it is only when someone asks me that I do not know it. One could say much the same about the flower, and perhaps about beauty itself, which like the flower quickly falls Time's victim. In fact, to this day every botanist who has wanted to define the flower has failed in the attempt. Let us now explain what this difficulty is, without, nevertheless, expecting any greater success than heretofore, if in my turn I try to wrestle with it.

Someone gives me a Rose, and he says, 'Here is a flower.'

That is showing it to me, I admit, but not defining it; and this inspection does not allow me to decide in the case of some other plant whether what I see is or is not a flower.

One generally thinks of the flower as the coloured part which forms the corolla, but one may easily be mistaken. There are bracts and also other organs which may be more colourful than the flower itself but which are not a part of it, as one sees in Cow-Wheats, and several Pigweeds and Goosefoots. Many flowers have no corolla at all; there are others which have a colourless one, so small and insignificant that only a most careful search can discover it. When Wheat is in flower, can you see the coloured petals? Can you see them in Mosses, in Grasses? Can you see them in the catkins of the Walnut, of the Beech and the Oak, in the Alder, the Hazel and the Pine, and in all those innumerable trees and herbaceous plants which only have stamens for their flowers? These flowers, nevertheless, are no less flowers. Thus, the essence of the flower lies not in its corolla.

This essence is not any clearer in any of the other parts which make up a flower, since there is not one of these parts which is not absent in some species of flowers. The calyx is missing, for example, in nearly all the Lily family; and one will not say that a Tulip or a Lily is not a flower. If there are some parts more important than others to a flower, they are undoubtedly the pistil and the stamens. Now in the whole of the Melon family, and indeed in all monoecious plants, half the flowers are without a pistil, the other half without stamens; yet this deprivation does not prevent them from being called and from being, each and every one of them, flowers. So the essence of the flower rests neither separately in any one of these so-called constituent parts, nor even in the aggregation of all these parts. In what, then, does this essence really lie? That is the question; that is the problem; and this is how Pontedera [Giulio Pontedera (1688-1757), botanist, author of *Anthologia Sive Floris Natura* (1720)] attempted to solve it.

The flower, he says, is a part of the plant which differs from the others in nature and in form; if it has a pistil, the flower is always attached to and of importance to the embryo; if it lacks a pistil, it is not attached to the embryo.

HAIRS SCALES TENDRILS (see Dictionary)

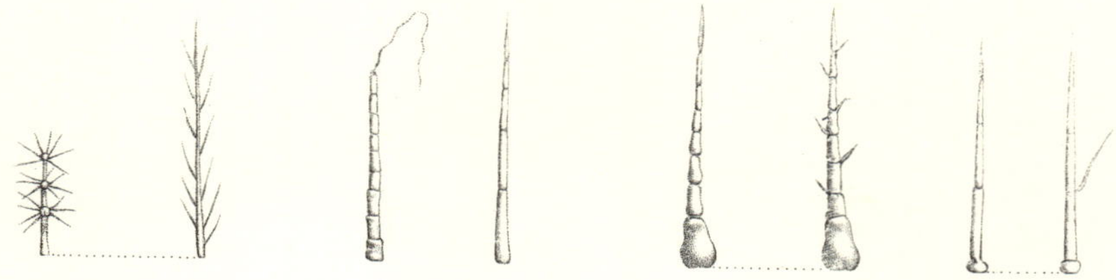

Hairs

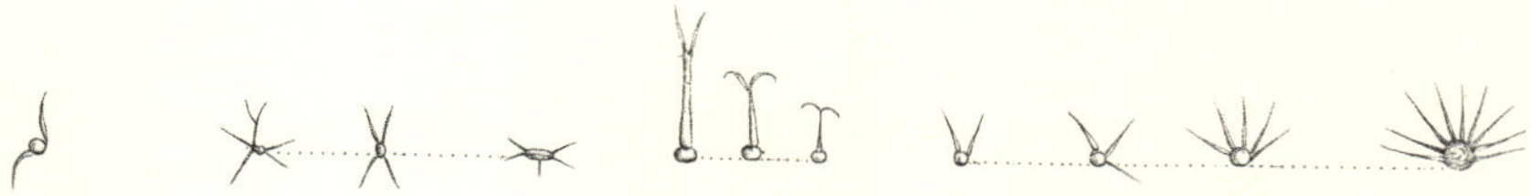

Hairs

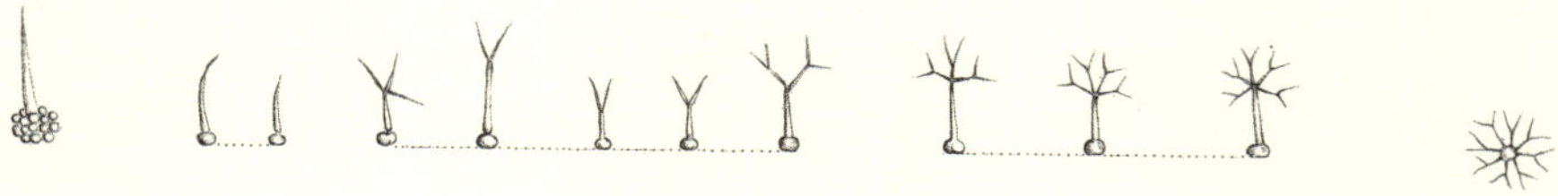

Hairs

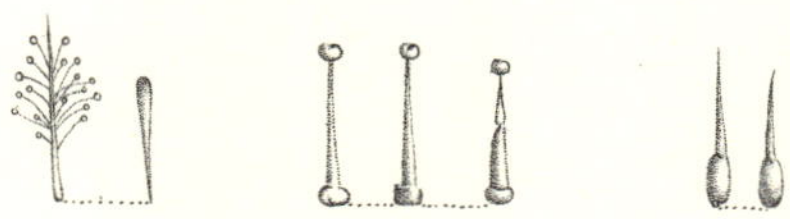

Scales

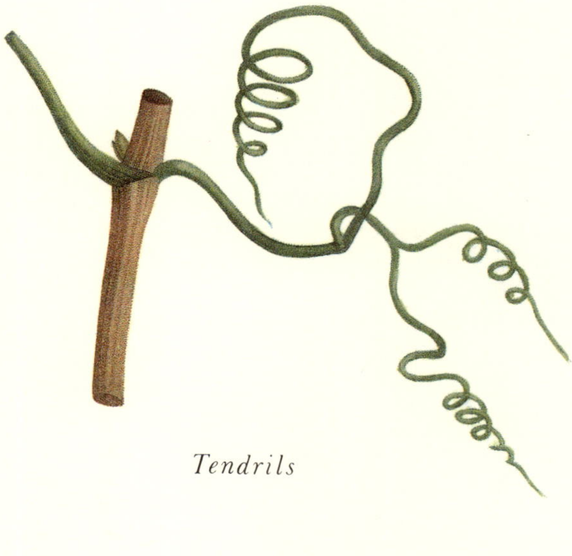

Tendrils

This definition falters, it seems to me, in that it is too all-embracing. For if there is no pistil, the only particular characteristic remaining to the flower is that it is different in nature and in shape from the other parts of the plant; one could, then, equally give the name to the bracts, or the stipules, to the nectary, to the thorns, and to everything which is neither leaf nor branch; and when the corolla has fallen and the fruit is nearing ripeness, one could still bestow the name 'flower' on the calyx and the receptacle, even though there would actually be no more flower left at that time. If, therefore, this definition fits *all*, it does not fit *individual instances*, and thereby falls short of one of the two requisite conditions. Besides, it leaves an emptiness in the spirit, which is the greatest fault a definition can have; for after having assigned the usefulness of the flower to the advantage of the embryo, when it is attached to it, it supposes the flower to be totally useless when it is not attached; and that fits poorly with the notion that botanists must have of the cooperation of the parts and of their role in the organic machine.

I think that the general shortcoming arises here from having considered the flower as an absolute, when it is only, it seems to me, a collective, relative entity; and from having too refined ideas, when one ought to restrict oneself to those which present themselves naturally. According to this reasoning, the flower seems to me to be only the transitory stage of the reproductive parts of the plant during fertilization of the seed. From this it follows that when all the reproductive parts are put together, there will only be one flower; when they are separated, there will be as many flowers as there are reproductive parts; and since these essential parts only number two, that is, the pistil and the stamens, there will in consequence be only two flowers, one male, the other female, which are necessary for reproduction. One can, however, imagine a third part which would reunite the sexes segregated in the other two. But then if all these flowers were equally fertile, the third would make the other two superfluous, and would alone be sufficient for the job; or there would really be two fertilizations, and we will now examine the flower in only one of them.

The flower, then, is merely the centre and instrument of fertilization. One on its own is self-sufficient only when it is hermaphrodite; when it is either male or female, there must be two flowers, that is, one of each sex. And if one considers in the make-up of the flower other parts, such as the calyx and the corolla, they cannot be essential parts, but merely the nutriments and preservers of those parts which are. There are even some with neither one nor the other. But there are none, and there could never be any, that at the same time have neither pistil nor stamens.

The flower is a localized and transitional part of the plant, which appears before the fertilization of the seed, and in which or through which fertilization takes place.

I shall not exert myself by justifying all the points of this definition here, for it is perhaps not worth the trouble. I shall only say that the phrase 'appears before' seems to me to be essential, because the corolla most frequently opens out and blooms before, in their own turn, the anthers open. And in this case, it is undeniable that the flower predates the process of fertilization. I add that this fertilization takes place *in* or *through* the flower, because in the male flowers of androgynous and dioecious plants no reproduction takes place, but they are none the less flowers for all that.

There, I believe, is the fairest assessment one can make of the flower, and it is the only one that leaves no room for objections which upset all the other definitions that have been attempted up to now.

As the flower is generally noticed for its corolla, a part much more striking than the others by virtue of the liveliness of its colours, it is also in this corolla that one automatically assumes the very being of the flower to dwell; and botanists themselves are not always free from this small error, for they often use the word flower for the corolla; but these little inadvertent inaccuracies

CRYPTOGAMS (see Dictionary under 'Cotyledon') *a* Epidermis of leaf *b* Leaves *c* Male flower *d* Peristome and cap of capsule *e* Spores *f* Cap, detached *g* Leaf tissue *h* Female flower *i* Isolated female flower *k* Complete plant (*a-k* all the same Moss, magnified) *l Lichen fraxineus* L.

a
b
c
d
e
f
g
h
i
k
l

are of minor importance, if they do not change one's view of things upon consideration. From the corolla stem the words monopetalous and polypetalous, labiate, personate, regular and irregular flowers, words which one frequently encounters even in scholarly books. This little inaccuracy was not only excusable but almost forced upon Tournefort and his contemporaries, who did not yet have the word corolla; and the usage has been preserved by custom since their days with no great inconvenience. But it would not be permissible for me, who point out this inexactitude, to reduplicate it here; thus I refer to the entry COROLLA for a discussion of its various shapes and parts. [The article on the COROLLA, to which the author refers here, was not found to have been written.]

But now I ought to discuss composite and simple flowers, for it is of the flower itself and not of the corolla that the flower consists, as we shall see after the exposition of the parts of the simple flower.

Simple flowers are divided into 'complete' and 'incomplete'. The complete flower has all the parts which are essential to or concomitant with reproduction, and there are four of these parts; two are essential, namely the pistil and the stamen or stamens; and two are accessory or concomitant, namely the corolla and the calyx, to which one should add the disk or receptacle which carries them all.

The flower is complete when it is composed of all these parts; when it lacks any of them it is incomplete. Now, the incomplete flower can lack not only the corolla and the calyx, but even the pistil or stamens; and in the latter case there is always another flower, either on the same plant or on another, which bears the other essential part lacking in the incomplete flower. As a result of this, the division can be made between hermaphrodite flowers, which may or may not be complete, and totally male or female flowers, which are always incomplete.

The incomplete hermaphrodite flower is no less perfect for that, since it is able to carry out fertilization by itself: but it cannot be called complete, since it is missing some one of the parts of those flowers which we call complete. A Rose and a Pink are, for example, perfect and complete flowers, for they are provided with all these parts; but a Tulip and a Lily are not complete flowers, though they are perfect, for they have no calyx. In the same way, the pretty little flower of the Pink family called Paronychia is perfect as a hermaphrodite; but it is incomplete, for despite its pleasant colour it lacks a corolla.

Without leaving this group of simple flowers I could now discuss regular flowers and flowers which are called irregular; but, as that distinction is concerned mainly with the corolla, it is better to refer the reader to that word. [COROLLA not included in the dictionary.] There are the contradictions posed by the name simple flower to be discussed.

When there is a single fruit, the whole flower is a simple flower. But if several fruits are produced by a single flower, that flower will be called composite; and this plurality of fruits is never to be found in flowers with only one corolla. [Incorrect. Simple flowers frequently produce several fruits.] Thus every composite flower must have not only several petals but several corollas; and for the flower to be a true composite, and not an aggregation of several simple flowers, one of the four parts must be common to all the florets which make it up, and must be missing on each individual floret. [There follows a discussion on composite flowers, the great part of which Rousseau covers in the Sixth Letter.]

* * * *

Simple flowers have another complication when they are called double or full.

The double flower is one whose parts are multiplied over and above its normal number, but without this multiplication being damaging to the germination of the seed.

Flowers rarely have double calyxes and practically never are the stamens increased. The commonest multiplication occurs in the corolla.

'ROOTS' *a* Strawberry runner *b* Fibrous root *c* Hyacinth bulb

a
b
c

The most frequent examples are found in polypetalous flowers such as Pinks, Anemones, Ranunculi. Monopetals double less frequently; however, one quite often sees double Campanulas, Primroses, Primulas and particularly Hyacinths.

The word double does not mean a simple duplication of the number of petals but some sort of multiplication. Whether the number of petals is doubled, trebled, or quadrupled, as long as they do not multiply to the point of suffocating germination, the flower is always called double; but when too many petals cause the stamens to disappear and the germ is aborted, then the flower loses the name double and is called fully double.

One sees from this that the double flower is still part of the natural order, but that the fully double flower is no longer part of it and is truly a monster.

Although the most common fullness in flowers is created by the petals, there are examples, nevertheless, where the calyx is involved, and we have a most remarkable example in the everlasting flower, *Xeranthemum*. This flower, which appears to be rayed, but which is really discoid, has like the Carline Thistle a calyx with overlapping bracts, whose inner row has long, coloured leaflets, which one would assume to be so many demi-florets, since they adorn the greater portion of the disk.

These false appearances often deceive the eyes of those who are not botanists; but anyone who has been initiated into the intimate structure of flowers cannot be deceived for a moment. A demi-floreted flower externally resembles a fully double flower; but there is always this essential difference – first, each demi-floret is a perfect flower which possesses its ovary, its pistil and its stamens; whereas in the fully double flower each multiplied petal is always simply a petal, which has none of the parts essential for reproduction. Take one after the other the petals of a single or double or fully double Ranunculus, and you will find nothing but the petal itself; but in the Dandelion each demi-floret, furnished with a style surrounded by stamens, is not just a petal but a true flower.

I am given the flower of a Yellow Water-Lily, and am asked if it is a composite flower or a double flower. I answer that it is neither one nor the other. It is not a composite since the petals which surround it are not demi-florets; and it is not a double flower, because doubling is not the natural state of a flower, and the natural state of the Yellow Water-Lily *is* to have several circles of petals round its ovary. Thus this multiplication of petals does not prevent the Yellow Water-Lily from being a simple flower.

The common state for most flowers is to be hermaphrodite; and this state would in effect appear to be the one most suited to the vegetable kingdom, where individuals, deprived of all spontaneous movement, cannot go looking for one another when the sexes are separate. In those trees and plants where the sexes are separate, nature, who knows how to vary her methods, has provided for this stumbling block; but it is no less true that, generally, stationary beings must, to perpetuate their species, have all the organs necessary for this end within themselves.

FRUCTIFICATION (Fructification) This term is always taken in a collective sense, and embraces not only the process of the fertilization of the seed and the ripening of the fruit, but also the aggregation of all the natural parts necessary for this operation.

FRUIT (Fruit) The end product of an individual plant's growth, containing the seeds which will regenerate the species through other individuals. The seed is only the end product when it is single and naked: when it is not so, it is merely a part of the fruit.

The word *fruit* has a much broader meaning in botany than in everyday use. In the case of trees and also some other plants all seeds or their edible coverings are given the general name *fruit*. But in botany the term refers even more broadly to all that transpires, after the flower has fallen, as a result of the fertilization of the seed. In this sense, the fruit is properly

COROLLAS *a* Dame's Violet *b* Tradescantia *c* Bindweed *d* Nightshade

nothing more than the fertilized ovary, and this is true whether it is edible or not, and whether the seed is ripe or unripe.

GENUS (Genre) Grouping of several species with a common characteristic which distinguishes them from all other plants.

GERM (Semence) The seed or simple rudiment of a young plant, which is combined with a substance necessary for its preservation before fertilization, and which feeds it during the first stages of germination until the plant can draw its nourishment directly from the soil.

GERM, EMBRYO, OVARY, FRUIT (Germe, Embryon, Ovaire, Fruit) These terms are so nearly synonymous that having discussed them in their separate articles I believe I should group them together here. The germ is the basic, primary element of the young plant; it becomes an embryo or ovary at the moment of fertilization, and this same embryo becomes a fruit on ripening: those are the exact differences. But one does not always observe these differences in usage, and often these words are employed indiscriminately.

There are two quite distinct types of germ, one contained in the seed, which as it develops becomes a plant, and the other contained in the flower, which through fertilization becomes a fruit. One can see the alternating process whereby each of these two germs generates itself and is in turn generated from it. [Germ is now only used as a synonym for embryo.]

Again, one can give the name germ to the rudimentary leaves enclosed in the leaf buds, and to the rudimentary flowers contained in the flower buds.

GERMINATION (Germination) Initial development of the parts of the miniature plant contained in the germ.

GLANDS (Glandes) Organs whose purpose is to produce plant secretions.

GRAFT (Greffe) An operation through which the juices of one tree are diverted through the vascular system of another; since the vascular systems of the two trees are not the same size or shape, nor are they butted exactly one against the other, the juices are forced to become refined as they separate, and this produces better and more flavourful fruit. [A fanciful description.]

GRAFT, TO (Greffer) To bind the shoot or bud of a healthy tree branch in the cortex of another tree, with necessary precautions and at the right season, so that the former receives the sap of the second tree and is nourished in the same way as it would have been by the tree from which it has been taken. The name *scion* is given to the piece that is inserted, and the word *stock* to the tree to which it is joined.

There are various methods of grafting. The approach graft, the cleft graft, the rind graft, the saddle graft, the bark graft.

GYMNOSPERM (Gymnosperme) A plant with seeds not enclosed in an ovary.

HAIRS or BRISTLES (Poils ou Soies) More or less solid, stiff threads growing from certain parts of plants. They are square or cylindrical, upright or lying flat, forked or simple, pointed or hooked; and these various forms are sufficiently constant to enable them to be used to classify these plants. [Hairs and bristles are not synonymous.]

HARDWOOD CUTTING – VINE (Maillet) Branch of new year's wood, on which, for replanting, two slivers of old wood are left protruding from each side. This type of cutting is used only for the vine, and even then quite rarely.

HEAD (Tête) A head-like or capitate flower is an aggregate or composite flowerhead, whose florets are arranged more or less spherically.

HERMAPHRODITES (Aphrodites) Monsieur Adanson [Michael Adanson (1727-1806), botanist, author of *Familles naturelles des plantes* (1763), who effectively refused to adopt the Linnean system] gives this name to animals among which each individual reproduces itself without any apparent act of copulation or fertilization. Examples are greenflies, molluscs, the majority of sexless worms, and those insects which reproduce themselves not through fertilization but by the division of a part of their body. In this sense, plants which are propagated by cuttings or by bulbils could be called hermaphrodite. This peculiarity, so contrary to the normal flow of nature, causes considerable problems in the definition of species. (Should one really say that in nature there are no species at all, only individuals?) But I believe one may doubt whether there are any completely hermaphrodite plants, that is to say ones which really do have no sex organs and cannot multiply by sexual reproduction. Also, there is a difference between the word *hermaphrodite* and *asexual*, for the first applies to plants which, having no sex organs, do not allow of multiplication, while the second refers only to those which are neuter or sterile, and are incapable of reproducing their own kind.

HOOD (Capuchon, Calyptra) A pointed cap which usually covers the capsule of Mosses. The hood is first attached to the capsule, but later it detaches itself and falls when the capsule nears maturity.

'HUSK', GLUME (Bale) Calyx-like bract of graminaceous plants (Grasses).

INFERIOR, SUPERIOR (Infère, Supère) There are two different positions of the calyx and corolla of flowers in respect of the seed, and the need to express this recurs so often, that one must create a word for it. When the calyx and the corolla are borne above the seed, the flower is said to be *superior*. When the seed is borne above the calyx and the corolla, the flower is said to be *inferior*. If the corolla is inferior then the seed is superior; if the corolla is superior, the seed is inferior: thus one has the choice of two ways in which to express the same thing. [In modern usage, the terminology is always expressed in relation to the seed (ovary). Thus when botanists today say *ovary superior*, they are speaking of the same arrangement as Rousseau's *superior flower*.]

As there are many more plants which bear inferior flowers than those with superior flowers, one should always assume the former condition, where it is not confirmed, since it is the more common; and if the description says nothing about the relative positions of corolla and seed, one must assume that the corolla is *inferior*: for if it were *superior*, the writer of the description would have expressly said so. [The priority Rousseau gives to inferior flowers does not hold.]

INTERNODES (Entre-Noeuds) These are the spaces which separate the nodes from which the leaves spring in grassy plants. There are some grasses, though very few, in which the stem is uninterrupted from one end to the other and has no nodes, and therefore no internodes. One such is Hair-Grass [*Molinia caerulea*]. [Incorrect. Hair-Grass has a single node hidden by leaves.]

KERNEL (Amande) Seed enclosed in a fruit-stone.

LAYER (Surgeon, *Surculus*) Name given to young stems of the Pink, etc., which one causes to take root by pressing them into the soil; they then produce another stem.

LEAF-BUD (Bourgeon) Early stage of leaves and branches.

LEAVES (Feuille) These are the organs necessary for plants to absorb humidity during the night and to facilitate transpiration during the day. They compensate for their lack of the spontaneous mobility of animals by allowing the wind to shake the plants and make them stronger. Alpine plants, ceaselessly assaulted by wind and storms, are always strong and vigorous; on the other hand, those raised in a garden have a delicate disposition, prosper less well, and often languish and degenerate. [Inaccurate and fanciful.]

LEGUMINOUS PLANTS (Legumineuses) See FLOWER.

LILIACEAE (Liliacées) Flowers with the characteristics of the Lily.

LIP (Limbe) When a regular monopetalous corolla widens and expands at its edges, the part which forms this widening is called the lip, and it is normally divided into four, five or more segments. Several Campanulas, Primulas, Bindweeds and other monopetals provide examples of this lip, which is to the corolla what the part we call the rim is to a bell. The different angle which the lip forms with the corolla tube gives the corolla the descriptions funnel-shaped, bell-shaped, cup-shaped.

MONOECIOUS (Monécie ou Monoecie) Plant shared by the two sexes. The name monoecious is given to a group of plants all of which have both male and female flowers on the same stock.

MONOECIOUS PLANTS (Monoiques) See above. Plants are called monoecious when they are not hermaphrodite but bear male and female flowers separately on the same stock. This word comes from the Greek, and means in this instance that the two sexes occupy the same house but not the same room. The Cucumber, the Melon and all the other plants of the Melon family are monoecious. [Later study makes Rousseau's definition inadequate: male and hermaphrodite, female and hermaphrodite, hermaphrodite and sterile flowers may all be found on the same stock and are termed monoecious. Furthermore, some plants of the melon family (e.g. Bryony) are dioecious.]

NAKED (Nu) Denied the coverings usually found in similar instances.

Seeds are called naked when they have no pericarp; umbels are naked when they have no involucre; stems are naked when they are not provided with leaves, etc.

NODES (Noeuds) The joints of stems and roots. [A node is the point of insertion of a leaf on a stem.]

OPPOSITE (Opposées) Opposite leaves are pairs of leaves ranged opposite each other on either side of the stem or the branches. Opposite leaves may be stalked or sessile. If there should be more than two leaves attached at the same level round the stem, then the term opposite is not used, and this arrangement of leaves takes on a different name. See WHORLED.

OVARY (Ovaire) This is the name given to the embryo of the fruit, or to the fruit itself before fertilization. After fertilization the ovary drops its name and is simply called fruit; or seed or spore if the plant is a gymnosperm. [The ovary is not the embryo, but the part of the pistil in which the seeds are contained. Gymnosperms, e.g. Conifers, have naked seeds and therefore no ovary.]

PALMATE (Palmée) A palmate leaf is one which, instead of being composed of several leaflets as is the digitate leaf, is merely divided into several lobes radiating from the top of the petiole but joined to each other at their point of attachment. [Palmate is nowadays synonymous with digitate. Other terms are used to describe other variations in the division of leaves.]

PANICLE (Panicule) A branched, pyramidal spike. The shape is produced because the lower branches, being the largest, create greater space between them; this becomes less higher up the stem, as the branches become shorter and fewer; in this way a perfectly regular panicle would terminate in a single sessile flower. [This is no longer correct. A panicle is a composite raceme, e.g. the Oat.]

PAPPUS (Aigrette) Tuft of simple or feathery filaments which top the seeds of several genera of Compositae and of some other flowers. The pappus may be sessile, i.e. attached directly to the seed which carries it, or pediculate, i.e. borne on a stem (called *Stipes* in Latin), which holds the pappus up above the seed. Initially the pappus serves as a calyx to the floret; later, as the floret withers, the pappus forces it away, so

SEX OF FLOWERS *a* Hermaphrodite flower: Periwinkle *b* Monoecious flowers: Castor Oil plant *c* Dioecious male flowers: Hemp *d* Dioecious female flowers: Hemp

a
b
c
d

that it is separated from the seed and does not hinder its ripening. The pappus protects the naked seed from rainwater, which could rot it, and when the seed is ripe, it acts as a wing by means of which the seed is carried away and scattered far and wide by the winds.

PARASITES (Parasites) Plants which are born or grow on other plants and feed on their substance. Dodder, Mistletoe and several Mosses and Lichens are parasitic plants. [Mosses and Lichens are not parasites but epiphytes.]

PARENCHYMA (Parenchyme) The pulpy substance or cellular tissue which forms the main substance of the leaf and the petal: in both cases it is covered by an epidermis.

PEDICEL (Pédicule) Elongated support which bears the fruit. [Modern usage has pedicel as the stalk of a single flower; peduncle as the stalk of an inflorescence.]

The adjective *pedicelled* seems to me necessary as the opposite of *sessile.* Botany is so surfeited with terms that one must make every effort to render those specially used in botany clear and brief.

The pedicel is the bond which attaches flower or fruit to branch or stalk. Its consistency is usually tougher than that of the fruit which is borne at one end, and less tough than that of the wood to which it is attached at the other. Usually, when the fruit is ripe, it detaches itself and falls with its pedicel. But sometimes, especially in herbaceous plants, the fruit falls and the pedicel remains, as one can see in the Dock family. One can also observe in this family another peculiarity. The pedicels, which all spring in whorls round the stem, are also all jointed at their centre. It seems in this instance that the fruit should detach itself at the joint, falling with one half of the pedicel and leaving the other half attached to the plant. However, that is not what happens. The fruit detaches itself and falls alone: the pedicel remains in one piece, and it requires a definite effort to divide it in two where it is jointed.

PERENNIAL (Vivace) Living for several years. Trees, shrubs, undershrubs are all perennials. Even several herbaceous plants are perennial, but only through their roots. Thus the Honeysuckle and the Hop are both perennial, though in different ways. The former retains its stems through the winter, so that they bud and flower in the following spring; but the Hop loses its stems at the end of each autumn, and starts to grow again every year from its rootstock.

Plants which are removed from their natural habitat are subject to variation in their perennial qualities. Several perennial plants of hot countries become annual with us, and this is not the only change they undergo in our gardens.

This means that exotic plants studied in Europe often give false impressions.

PERFOLIATE (Perfoliée) The perfoliate leaf is that in which the stalk passes through its centre and is completely surrounded by the leaf.

PERIANTH (Périanthe) A type of calyx which closely envelops the flower or the fruit. [Today, perianth refers to calyx and corolla together.]

PETAL (Pétale) The name petal is given to each complete piece of the corolla. When the corolla consists of only one piece, there is also only one petal; in this case the petal and the corolla are one and the same thing, and this sort of corolla is termed monopetalous [gamopetalous]. When the corolla is made of several pieces, these pieces are so many petals, and the corolla which they make up is described by the Greek word for the number of the petals, because the word petal also comes from the Greek, and it is seemly, when one wants to create a word, to draw its component roots from the same language. Thus the words monopetal, dipetal, tripetal, tetrapetal, pentapetal, and finally polypetal, apply to a corolla of one piece, or two, or three, or four, or five, etc., and finally of an indeterminate number of pieces.

PETALOID (Pétaloïde) Having petals. Thus a *petaloid* flower is the opposite of the *apetaloid* flower. [Rousseau has confused his terms. Petaloid means 'like a petal'.]

Sometimes this word is used as the second part of a word whose first part designates a number. Then it means a monopetal deeply divided into as many sections as are indicated in the first part of the word. Thus the tripetaloid corolla is divided into three parts or half-petals, the pentapetaloid corolla into five, etc.

PETIOLE (Pétiole) The elongated support which bears the leaf. The word *petiolated* is the opposite of *sessile* with regard to leaves, just as the word *pediculated* is in regard to flowers and fruits. See PEDICEL, SESSILE.

PINNATE (Pinnée) A leaf bordered by several leaflets is called a pinnate leaf.

PISTIL (Pistil) Female organ of the flower, surmounting the germ, and through which the seed receives the fertilizing pollen from the anthers. The pistil is normally lengthened by one or several styles; it is also sometimes directly surmounted by one or several stigmas, with no intervening style. The stigma receives the abundant pollen from the top of the stamens, and passes it through the pistil into the centre of the germ to germinate the ovary [ovule]. Following the sexual system, plant fertilization can only be carried out by the coming together of the two sexes, and the act of fruiting is simply that of regeneration. The filaments of the stamens are the *vas deferens*, the anthers the testicles, the pollen which they disseminate is the seminal fluid, the stigma becomes the vulva, the style is the vagina, and the germ does the work of the uterus or womb.

PLACENTA (Placenta) The part of the carpel which bears the ovules. The ovules (seeds) are directly attached to the placenta. Linnaeus does not allow this word *placenta*, and always uses *receptacle*. However, these words have very different meanings. The receptacle is the part by which the seeds are attached to the carpel. It is true that when the seeds are naked, there is no placenta other than the receptacle; but in all angiosperms, the receptacle and the placenta are different.

The septa of all ovaries with several carpels are true placentas, and in single celled ovaries the ovary is the only placenta.

PLANT (Plante) A vegetable body made up of two principal parts, namely, the root by which it is anchored in the ground or to another body from which it draws its nourishment, and the part above the ground, through which it inhales and breathes out the element in which it lives. Of all known vegetables, the Truffle is almost the only one which one could not describe as a plant.

PLANTS (Plantes) Vegetable matter spread over the face of the earth to clothe it and beautify it. There is no sight so sad as that of bare soil; there is no sight so joyful as that of mountains clad with trees, rivers lined with thickets, fields carpeted with green, and valleys sprinkled with flowers.

POD (i) (Cosse) Pericarp of leguminous plants. The pod is usually composed of two valves, and sometimes only one.

(ii) (Légume) A type of pericarp made up of two valves joined the length of their edges by two longitudinal sutures. The seeds are attached by the upper suture to the two valves alternately; the lower suture has no seeds. The word legume (pod) is the normal one for the fruits of leguminous (papilionaceous) plants.

POD, SHELL, HUSK (Gousse) Fruit of a leguminous [papilionaceous] plant. The pod is also called the legume, and is usually composed of two panels called valves which may be flat or convex and are tightly connected by two longitudinal sutures. They enclose the seeds, which are alternately attached to one or other valve at the suture, and detach themselves when they are ripe.

POLLEN (Poussière prolifique) This is a mass of tiny spherical particles enclosed in each anther which, when the anther opens and releases them on to the stigma, in turn open, and permeate the stigma with a fluid which penetrates down through the pistil and fertilizes the fruiting embryo.

POLYGAMY (Polygamie) Polygamous plants embrace all those which have hermaphrodite flowers on one stock and single sex flowers, male or female, on another.

The word polygamy is also applied to several

divisions of composite flowers, and in this case the meaning is slightly different.

All composite flowers may be regarded as polygamous, since they all comprise several florets which germinate separately and which therefore each have their own place and, one might say, their own family line. All these separate units are connected in various ways, and because of this they form various combinations.

When all the florets of a composite are hermaphrodite, they form a system named regular polygamy.

When all these florets are not hermaphrodite, they form, as it were, a system named bastard polygamy. This has several divisions.

1 *Superfluous polygamy*, when the disk florets, all being hermaphrodite, germinate, and the female ray florets also germinate.
2 *Useless polygamy*, when the disk florets, being hermaphrodite, germinate, while the ray florets are neuter and do not germinate.
3 *Essential polygamy*, when the disk florets are male and the ray florets are female, they need each other to germinate.
4 *Separated polygamy*, when the component florets are split up in ones or several together by a number of partial calyxes contained within the calyx of the complete flower.

One could make up more new combinations, by imagining, for example, male ray florets, and hermaphrodite or female disk florets; but this does not occur.

[The above system is not in use today.]

PULP (Pulpe) The soft, fleshy matter of many fruits and roots.

RACEME (Grappe, *Racemus*) A type of spike in which the flowers are neither sessile nor directly attached to the central stem, but to secondary pedicels into which the primary pedicels are divided. The raceme is merely a panicle whose branches are denser and shorter, and often thicker than in the true panicle.

When a panicle or spike hangs downwards, rather than raising itself towards the sky, it is called a *raceme*; this is the case with the red-currant, and with a bunch of grapes. [A panicle is a branched raceme. Racemes do not always hang downwards.]

RADICAL (Radicales) Applied to the leaves which are closest to the root: the word is also extended to stems in the same sense.

RECEPTACLE (Réceptacle) That part of the flower and fruit which forms a base for all the others, and through which the plant supplies to them the nutritious juices which it must extract.

It is usually divided into the receptacle proper, which bears only a single flower and a single fruit, and consequently is found only in the simplest flowers; and the communal receptacle, which carries several flowers.

When the flower is inferior, the same receptacle supports the whole inflorescence; but when the flower is superior, the receptacle proper is a double one, and that which bears the flower is not the same as that which bears the fruit. This is true in the most usual arrangement; but on this subject one can pose the following problem, for whose solution nature has created one of her most ingenious inventions.

When the flower is above the fruit, how can it happen that the flower and the fruit have one and the same receptacle?

The communal receptacle properly belongs only to composite flowers, whose florets it bears and unites into one, regular flower, so that to cut out some of the florets would make the whole irregular; but apart from aggregate flowers about which one can say much the same thing, there are other types of communal receptacle which are still entitled to the same name since they serve the same purpose. Such are the *umbel*, the *spike*, the *panicle*, the *thyrsus* [a contracted panicle], the *cyme*, the *spadix*, about which you will find entries in the appropriate places.

REGULAR FLOWERS (Fleurs Régulières) Regular flowers are symmetrical in all their components, as in the Crucifers, Lilies, etc. [i.e. actinomorphic as opposed to zygomorphic or irregular].

STEMS *a* Rose bush *b* Cistus *c* Oak

a
b
c

RENIFORM (Réniforme) Kidney-shaped.

ROOT (Racine) Part of the plant by which it is held in the soil or to the body which nourishes it. Plants, attached in this way, cannot move their position; feelings would be useless to them, since they cannot go in search of their needs, nor can they flee from what is harmful: nature creates nothing superfluous.

ROSACEOUS (Rosacée) A regular polypetal, like the Rose.

ROSETTE (Rosette) A monopetalous flower with no corolla tube, or a very short one, and a much flattened rim. [Used now only to describe a rose-like arrangement of leaves.]

SAP (Suc nourricier) Portion of the plant juices which feeds the plant.

SCALES (Ecailles ou Paillettes) Small, scaly tongues, which in several of the Composite genera are embedded in the receptacle, and mark out and separate the florets. When the scales are merely thread-like they are called 'down'; but when they are somewhat more sturdy they are called 'scales' ['bracts'].

It is peculiar to the double everlasting flower Xeranthemum that the scales round the disk grow longer, take on colour and appear to be genuine demi-florets, to the point of deceiving the quick glance of someone who does not peer more closely.

Very commonly the word scales is given to the calyxes of catkins and fir-cones. The name is also given to the overlapping leaflets of the calyxes of flowerheads such as Thistles, Knapweeds, and to those with dry, membraneous calyxes such as the Xeranthemum and Cupidone.

In some species, the stem is also covered with scales. These are leathery, rudimentary leaves which sometimes take the place of true leaves in such plants as Broomrapes and the Coltsfoot.

Finally, the overlapping coverings of the bulbs of several Lilies, and the flattened calyxes of the Bog-Rush and some of the Grasses are called scales.

SCAPE (Hampe) A leafless stem, specially designed to keep the fruiting parts high above the roots. [A scape may sometimes have one or more leaves, all radical, e.g. Tulip.]

SESSILE (Sessile) This adjective denotes the absence of a receptacle. It means that the leaf, the flower or the fruit to which the word is applied is directly attached to the plant without an intermediate petiole or pedicel. [Rousseau is not using the word receptacle here in the sense in which he defines it in this Dictionary.]

SEX (Sexe) This word has been extended to the vegetable kingdom, and has become a familiar one since the establishment of the sexual system.

SHEATH (Enveloppe) A type of calyx which surrounds flowers, as in the Arum, the Fig, and flowers with florets. Flowers with sheaths are not necessarily without true calyxes.

SHRUB (Arbrisseau) Woody plant smaller than a tree. It usually branches from the base into several stems. Trees and shrubs produce buds in autumn in the leaf axils. In the spring these develop into flowers and fruit – a characteristic which distinguishes them from undershrubs.

SILIQUA (Silique) Fruit composed of two valves, held together by longitudinal sutures to which the seeds are attached on each side.

Normally the siliqua is divided into two cells (bilocular), and partitioned by a septum to which are attached a number of the seeds. However, this septum, since it is not an essential part, should not be included in the definition, as one can perceive in the Caper, Greater Celandine, etc.

SOLITARY (Solitaire) A solitary flower is one which grows alone on its pedicel.

SPADIX or FLESHY SPIKE (Spadix ou Régime) This is the flower spike of the Palm family. It is really the fruiting receptacle, and it is surrounded by a spathe, which acts as a cloak.

SPATHE (Spathe) A sort of membranous calyx [involucral bract] which acts as a covering for the flowers before they open, and which splits apart to allow a free passage when fertilization is about to take place.

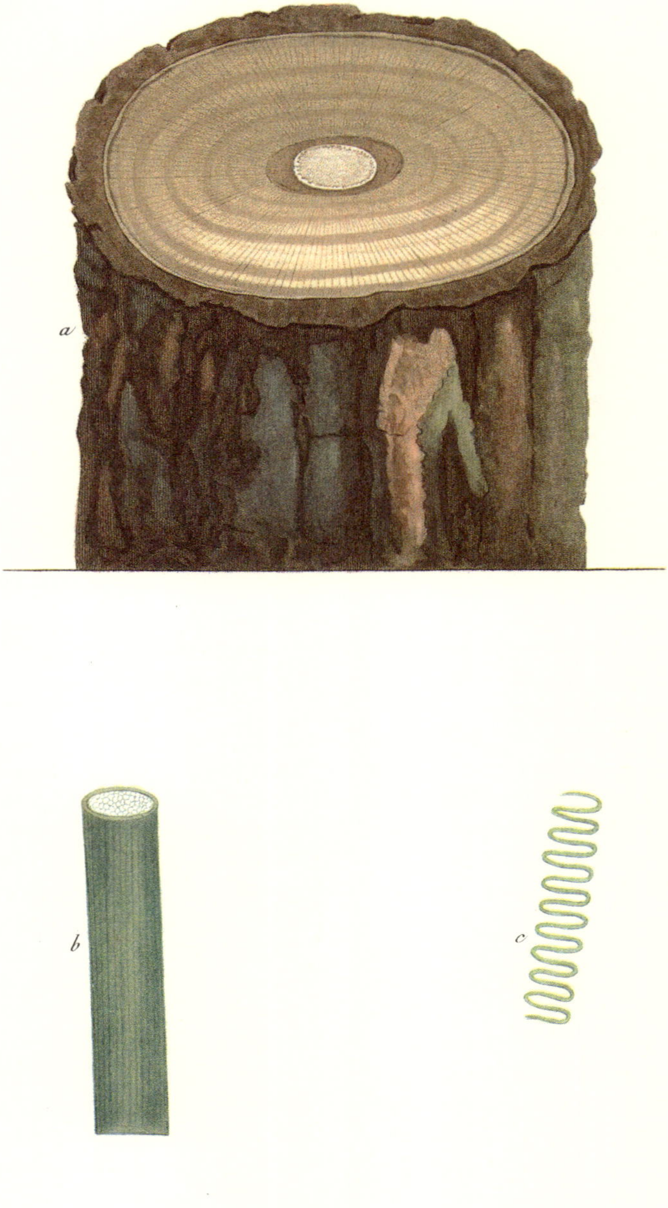
a
b
c

The spathe is characteristic of Palms and Lilies.

SPECIES (Espèce) The grouping of several varieties or individuals under a common characteristic which distinguishes them from all other plants of the same genus.

SPUR (Éperon) Conical protuberance, straight or curved, formed in several flowers by the lengthening of the nectary. Examples are the spurs of Orchids, of Toadflaxes, of Aquilegia, of Larkspurs, of several Geraniums and of other flowers. [It is not the nectaries but parts of the calyx or corolla which are spurred.]

STAMENS (Étamines) The masculine agents of fertilization. Their form is usually that of a filament which supports a head called an anther. This anther is a kind of capsule containing a mass of pollen. The pollen is released either explosively or by dispersion, and penetrates the stigma, to be borne to the ovaries which it then fertilizes. Stamens vary both in shape and number.

STANDARD (Etendart) The upper petal of leguminous flowers [Peas, etc.].

STEM (Tige) The trunk of the plant, from which all the other parts above the ground arise. It has affinities with the midrib, in that it is sometimes solitary and branches in the same way – for example in the Fern; but it differs in being symmetrical and in having no front or back or obvious sides, all of which features are found in the midrib.

Several plants have no stem, others have only a naked, leafless stem, which in this case is called by another name. See SCAPE.

STIGMA (Stigmate) Tip of the pistil which becomes moist at the moment of fertilization, so that the pollen adheres to it.

STIPULE (Stipule) A type of leaflet or scale which grows at the base of the petiole, the pedicel or the branch. Stipules usually lie outside the part they are accompanying, and act as a kind of bracket, but they also sometimes grow beside, opposite or even inside the axil.

Monsieur Adanson states that the only true stipules are those which are attached to stems, as in Bilberries, Periwinkles, Spurges, Buckthorns, Chestnuts, Limes, Mallows, and Capers: they take the place of leaves in plants where they are not in whorls. In leguminous plants the position of the stipules varies. In the Rose family there are no true stipules, merely a prolongation or appendage of the leaf, or an extension of the petiole. There are also membranous stipules, as in the Spurreys.

STOLON (Traînasse ou Traînée) A long stem which in certain plants creeps along the ground and at intervals has joints from which small roots penetrate the ground and produce new plants.

STONE (Noyau) [Endocarp] Bone-like shell which encloses a kernel.

STYLE (Style) The part of the pistil which keeps the stigma above the ovary. [A style may be lacking, e.g. in the Poppy.]

SUCKERS (Drageons) Stems which root and spring from the base of a tree, or from the trunk, which cannot be pulled out without damaging the tree.

SYNONYM (Synonymie) List of various names given by different authors to the same plants.

Synonomy is by no means an idle and pointless study.

TENDRILS (Vrilles ou Mains) Thread-like organs which terminate the branches of some plants and provide them with the means of attaching themselves to other bodies. Tendrils are simple or branched; being unfettered, they go in all sorts of directions; and when they encounter a foreign body they wind round it spirally. [Rousseau is only incorrect in limiting tendrils to extensions of branches. They may also spring from the leaves, the roots or the axes, depending on the species.]

TERMINAL (Terminal) A terminal flower is one borne at the apex of the stem or of a branch.

TERNATE (Ternée) A ternate leaf is composed of three leaflets attached to the same petiole.

THYRSUS (Thyrse) A branched, cylindrical spike [contracted panicle]. This term is seldom

STRUCTURE OF STEMS *a* Grass *b* Pink *c* Dandelion

a
b
c

used, as examples are not common. [The Lilac and the Horse Chestnut are examples.]

UMBEL (Ombelle) An arrangement of spokes, which arise from the same point, and spread out like the spokes of a parasol. The umbel is carried at the end of the stem or branch; the secondary umbel arises from a spoke of the umbel [i.e. it is a small umbel forming part of a composite umbel].

UNDERSHRUB (Sous-arbrisseau) A woody plant or small bush, not as large as a shrub, which does not produce flower or fruit buds in the autumn. For instance, Thyme, Rosemary, Gooseberry, Heathers, etc. [An undershrub is not differentiated by bud production from shrubs, but only by size.]

VEGETABLE (Végétal) Organic body endowed with life and deprived of feeling.

They will not let me off with that definition, I know. People want minerals to live, vegetables to feel, and even formless matter to be imbued with feeling. Although that may be so according to modern physics, I have never been able, nor will I ever be able to express the ideas of other people, when these ideas do not coincide with mine. I have often seen a dead tree which I previously saw alive, but the death of a stone is an idea which would never enter my head. I see delicate feeling in my dog, but I have never noticed any in a Cabbage. Jean-Jacques's paradoxes are well known. I dare wonder if he ever put forward any as silly as that with which I would have to contend if I now entered on this discussion. Yet nobody is shocked by it. Enough; I will return to the subject.

Since vegetables come into being and live, they must also destroy themselves and die; it is the irrevocable law to which all organisms are subject: in consequence, they reproduce. But how does this reproduction come about? We see everything in the vegetable kingdom which is submitted to our perceptions created through reproduction; and we can presume that this natural law is equally obeyed in those parts of the same kingdom whose methods escape our eyes. I see neither flowers nor fruits in Silkweed, in scum-forming algae, in Truffles; but I see these vegetables perpetuating themselves; and by analogy we can attribute to them the same methods of fertilization as the other vegetables to attain the same end; this analogy seems so true that I cannot deny it my assent.

It is true that the majority of plants have other means of perpetuating themselves, such as bulbils, cuttings and rooted suckers. But these methods are secondary rather than basic: they are by no means common to all plants; only fruit-producing is, and since there are no exceptions to this principle in those plants well known to us, one cannot suppose that it is any less universal in other vegetable organisms.

VINE STOCK (Provin) A vine stem bent over and pegged into the earth. It produces hairy roots at its nodes which grow into the soil. One then cuts the stem holding it to the stock, and the other end which is growing in the ground becomes a new stock.

WATER CHANNELS, GUTTERS (Abreuvoirs ou Gouttières) Formed in the wood of rotten tree stumps, they rot the rest of the trunk by retaining rainwater.

WHORLED (Verticillé) More than two organs arising at the same level round a common axis.

WOODY (Ligneux) Having the consistency of wood.

LIST OF REDOUTÉ ILLUSTRATIONS

BIBLIOGRAPHY

Blunt, Wilfred, *The Art of Botanical Illustration*, Collins, London, 1950

Dunthorne, Gordon, *Flower and Fruit Prints of the Eighteenth and Early Nineteenth Centuries*, Dulau & Co., London, 1939

Léger, Charles, *Redouté et son temps*, Edition de la Galerie Charpentier, Paris, 1945

Plesch, Arpad, *Mille et un livres botaniques*, Arcade, Brussels, 1973

Rousseau, Jean-Jacques, *Confessions*, Editions Garnier Frères, Paris, 1964

Rousseau, Jean-Jacques, *Les rêveries du promeneur solitaire*, Garnier-Flammarion, Paris, 1964

Rousseau, Jean-Jacques, *Lettres sur la botanique*, Introduction by Bernard Gagnebin, Club des Libraires de France, Paris, 1962

Rousseau, *Oeuvres complètes*, Vol. IV, Introduction to *Lettres sur la botanique* et *Fragments pour un dictionnaire de botanique* by Roger de Vilmorin, Bibliothèque de la Pléiade, 1969

Sénelier, Jean, *Bibliographie générale des oeuvres de J.-J. Rousseau*, Presses Universitaires de France, Paris, 1950

Sitwell, Sacheverell and Wilfred Blunt, *Great Flower Books, 1700-1900*, Bibliography edited by Patrick M. Synge, Collins, London, 1956

The whereabouts of Redouté's original paintings for the *Botanique* are unknown. Hundreds of his water-colours on vellum are preserved, however, in the Muséum National d'Histoire Naturelle in Paris.

Herbaria created by Rousseau, or parts of them, survive in Paris at the Musée Carnavalet, the Musée des Arts Décoratifs, and the Muséum National d'Histoire Naturelle; and at the Musée de Chaalis near Ermenonville, and the Central Library in Zurich. One of his finest, in eleven volumes, was destroyed in Berlin during the Second World War.

The publishers would like to thank Ruth Schneebeli-Graf for her help and advice.